Carl Douglas

J. Reinert M. M. Yeoman · Plant Cell and Tissue Culture

J. Reinert M. M. Yeoman

Plant Cell and Tissue Culture

A Laboratory Manual

Layout and illustrations
by P. Macdonald

With 37 Figures

Springer-Verlag
Berlin Heidelberg New York 1982

Prof. Jakob Reinert,
Institut für Pflanzenphysiologie und Zellbiologie,
Königin-Luise-Straße 12–16a,
1000 Berlin 33, Federal Republic of Germany

Prof. Michael Magson Yeoman,
University of Edinburgh,
Department of Botany,
The King's Buildings, Mayfield Road,
Edinburgh EH9 3JH, Great Britain

Patricia Macdonald
Graphic Design and Illustration
6, Ashgrove
Musselburgh, East Lothian EH21 7LT, Great Britain

ISBN 3-540-11316-9 Springer-Verlag Berlin Heidelberg New York
ISBN 0-387-11316-9 Springer-Verlag New York Heidelberg Berlin

Printed in Germany.

Typesetting, printing, and bookbinding by Brühlsche Universitätsdruckerei Giessen
2131/3130-543210

Contents

I Introduction

The techniques of plant organ, tissue, and cell culture are now established in many research laboratories throughout the world and are being used in numerous areas of plant science. Methods have been developed to propagate plants and free them from viruses using shoot tip culture. The regeneration of plants from callus culture has also proved useful commercially. Elegant techniques have been used to synthesise somatic hybrids by the fusion of protoplasts and to transform cells. These and many other techniques have been and can be used to investigate a variety of botanical phenomena as well as to improve crop plants and now provide an important part of the basic experimental skills required by a majority of experimental botanists. Therefore the teaching of the principles and methods of plant organ, tissue, and cell culture is an essential part of the training programme of undergraduates and technicians. There are now several excellent texts covering the basic theoretical material of the subject and the application of the techniques to a variety of applied situations in agriculture, forestry and horticulture. However, much of the simple practical details can only be acquired by attending a course or working alongside an experienced laboratory worker; this is not always possible and many wish to discover the basic practical fundamentals of the subject inexpensively and without travelling around or carrying out an extensive research of the literature. In short, there is a necessity for a basic laboratory manual in which the complete details of materials and equipment together with full instructions of how to perform each manipulation successfully, are presented clearly and without ambiguity within the different areas already developed. Additionally it is necessary to provide, as far as possible, techniques which can be performed by a majority of relatively inexperienced students and which can be guaranteed to succeed at least nine times out of ten. The experiments included in this volume have been performed many times in undergraduate courses within the laboratories of the authors and comprise an essential collection which can be carried out with the minimum of facilities. There is no pretence of sophistication in this manual; only enough theoretical background is included to put the experiment into context. Mainly general references have been included to direct the reader to review articles in text books. We have concentrated on reproducibility, simplicity and accuracy with sufficient illustration to make all manipulations clear.

The drawings of items used in the bench layout diagrams are symbolic and are 'keyed in' by number to the list of materials and equipment. A line around an item indicates that is sterile.

The adoption of an integrated text in which diagrams are related spatially to the methods will, we hope, help the student to grasp the techniques quickly and effectively. This is first and foremost a manual which has its place on the laboratory bench open in front of the student, a book to be used!

The experiments are grouped together under appropriate headings and when taken together in sequence form a comprehensive course on the techniques of organ, tissue, and cell culture. A shorter course may be devised by selecting one experiment from each group, or where desired, all of the experiments within one group may be chosen to emphasise a particular area of study. Some of the experiments may be performed by a single student but the optimal size of the working group for university undergraduates is two. The Appendix provides details of culture media, materials, and various methods.

We are delighted to acknowledge all of our students and colleagues past and present who have assisted in the moulding and improvement of the experiments in this manual. We would especially like to thank Dr. Pat Macdonald who designed the basic format and provided all of the excellent illustrations.

II Isolation of Plant Material and Studies on Growth and Cell Division

II Isolation of Plant Material and Studies on Growth and Cell Division

Experiment 1 Isolation of Explants, Establishment and Maintenance of Callus *(Daucus carota)*

Explants isolated from the tissues of higher plants and brought into culture, like excised organs, require a nutrient medium consisting of a mineral salts mixture, a carbon source, (usually sucrose) and vitamins. In addition phytohormones (auxins and cytokinins), or their synthetic counterparts, are required to initiate and maintain cell division; occasionally other organic supplements, for instance amino acids or hexitols, are necessary to ensure the prolonged growth of the excised tissue to give an established callus. On suitable media, tissue fragments from most dicotyledonous plants, particularly those containing meristematic cells, (for instance the cambium or shoot meristem), will start to proliferate rapidly. Less frequently some of the mature cells will also divide and subsequently become involved in callus formation. Objects best suited for such experiments are explants cut from the tap root of carrot *(Daucus carota)*. These may also be used to test the effects of different compounds of the culture medium on induction and subsequent growth of callus from explants.

Materials and Equipment
Sterile Items
(For sterilisation procedures and composition of medium see Appendix)

1 12 rimless culture tubes (150 × 25 mm) containing 10 ml of 0.8% agar medium $M_{DAUC} + 5 \times 10^{-8}$ g/ml 2,4-D sloped at an angle of 25°
2 20 sheets of tissue paper (200 × 200 mm)
3 10 Petri dishes 90 mm in diameter, either glass or plastic
4 3 l of sterile distilled water contained in 3 Erlenmeyer conical flasks (1,000 ml)
5 50 sheets of aluminium foil (100 × 100 mm)
6 2 crystallising dishes with lids (90 mm in diameter and 50 mm deep)
7 2 pairs of forceps (120–150 mm)
8 3 scalpels (c. 150 mm)

Non-Sterile Items

9 2 tap roots of carrot *(Daucus carota)* at least 200 mm in length and 40 mm in diameter. The plant material may be purchased at the local

Fig. 1.1 Items for the sterile transfer room

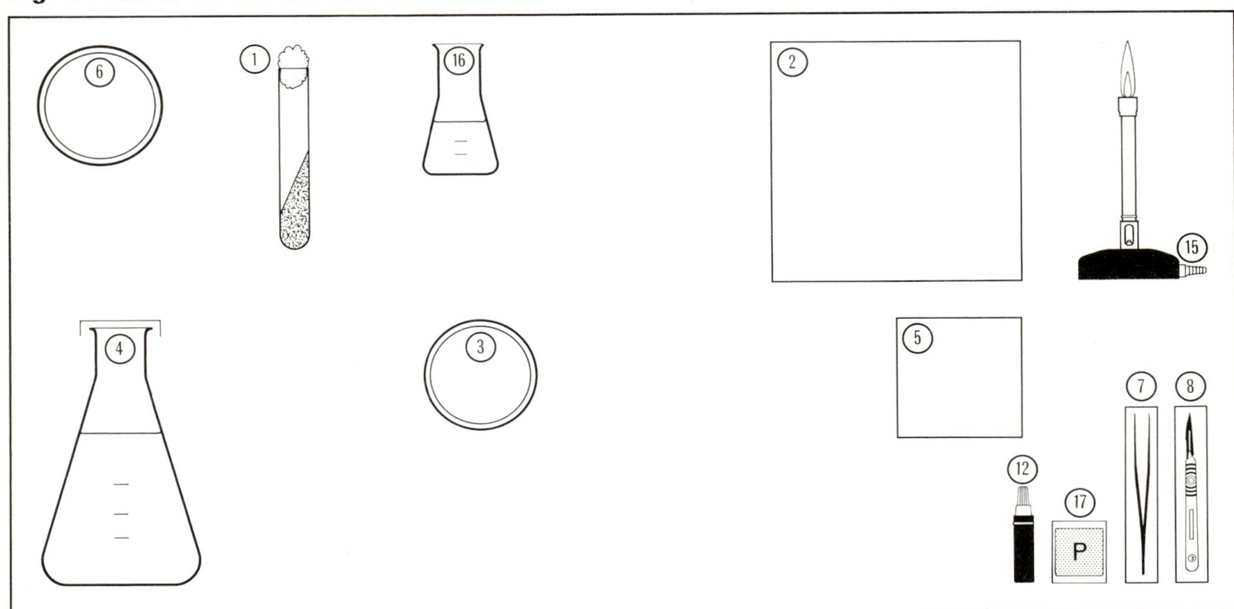

market or preferably grown in open ground, in sandy soil in the garden. Most varieties will respond to the culture procedure employed in this experiment.

10 2 racks, preferably plastic or metal to hold 12 culture tubes (150×25 mm) at an angle of approximately 25°

11 1,000 ml of a solution of sodium hypochlorite approximately 20% (v/v), commercial bleach preparations such as Ace or Domestos can be used at a dilution of one part to five of water

12 1 waterproof marking pen

13 1 glass beaker (1,000 ml)

14 1 analytical balance for determining the fresh weight of the cultures

15 1 Bunsen or ethanol burner

16 1 Erlenmeyer flask (150 ml) containing 100 ml 95% ethanol

17 1 roll of parafilm

18 1 small nylon nailbrush

The items for the sterile transfer room are laid out as shown in Fig. 1.1.

Experimental Procedures

Reject all diseased, damaged, or irregularly shaped individuals. Scrub the carrots with a small nylon nailbrush under running tap water to remove all surface detritus. Place the carrots (trimmed to c. 100 mm in length) in a 1,000 ml beaker, cover with the solution of sodium hypochlorite and leave for approximately 30 min. During the sterilisation procedure mark each culture tube with the number of the carrot root, the medium and the date. Transfer sterilised carrots to the sterile room making sure the UV lights are switched off before entering. Powerful UV rays are harmful to the eyes and skin! After wiping it clean with 70% ethanol set out the working table according to Fig. 1.1. Throughout the manipulation sequence forceps, scalpels, and other small instruments must be kept in 95% ethanol and flamed thoroughly before use. Ethanol is inflammable, take great care during flaming! Wash the carrots three times with sterile distilled water agitating the beaker to completely remove the hypochlorite, and dry using the tissue paper. Transfer a carrot with 20 mm removed from each end to a sterile crystallising dish and cut a series of transverse slices 1 mm in thickness from the carrot root, using a sharp scalpel (Fig. 1.2A). Transfer each slice to a sterile Petri dish and cut explants 4 mm×4 mm square across the cambium so that each piece contains parts of the phloem, cambium and xylem (Fig. 1.2B). Size and thickness of the explants should be uniform, so that a comparison can be made of the fresh-weight of the cultures after 4 weeks of growth. (Weigh 10 freshly isolated explants in order to determine the average fresh-weight.) Repeat this procedure until sufficient explants have been accumulated for the experiment. Always replace the lid of the Petri dish after each manipulation. Remove the closure from a culture tube and flame the uppermost 20 mm of the open end. while holding the tube at an angle of 45°, transfer (using forceps) two explants onto the surface of the agar, taking care that the root pole is touching the medium. Seal each tube immediately with a square of aluminium foil which has been flamed, before and after it has been placed on the tube. Always flame the forceps between transfers. Repeat this procedure until all of the culture tubes have been used. Place the culture tubes in the racks and incubate in the dark at 25 °C.

Fig. 1.2 Isolation of explants

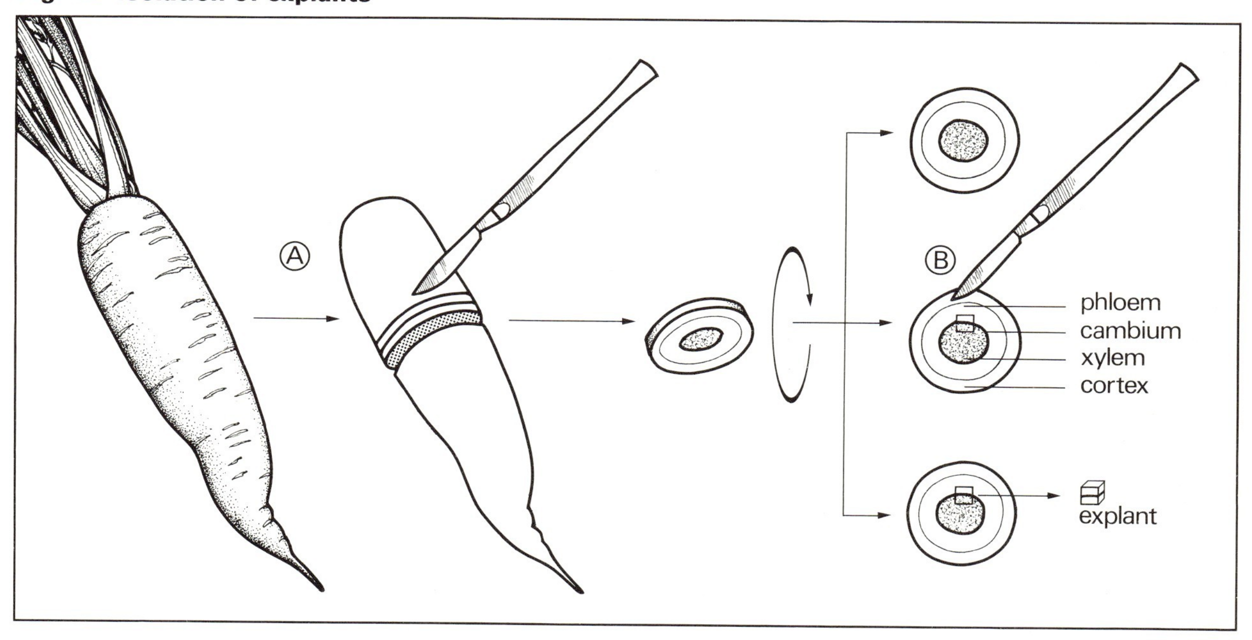

Sub-Culture of Callus

Materials and Equipment

Sterile Items

(For sterilisation procedures and composition of medium see Appendix)

1 5 carrot cultures initiated from explants 4–6 weeks ago *or* an established callus sub-cultured 3–4 weeks previously
2 12 rimless culture tubes (150×25 mm) containing 10 ml of 0.8% agar medium $M_{DAUC}+5\times10^{-8}$ g/ml 2,4-D sloped at an angle of 25°
3 20 sheets tissue paper (200×200 mm)
4 5 Petri dishes, 90 mm in diameter
5 50 sheets aluminium foil (100×100 mm)
6 3 pairs forceps (120–150 mm)
7 3 scalpels (150 mm)

Non-Sterile Items

8 2 racks, preferably plastic or metal, each to hold 12 culture tubes (150×25 mm)
9 1 waterproof marking pen
10 1 Erlenmeyer flask (150 ml) containing 100 ml 95% ethanol
11 1 roll parafilm
12 1 Bunsen or ethanol burner

The items for the sterile transfer room are laid out as shown in Fig. 1.3.

Experimental Procedures

Remove the culture tubes one at a time from the rack and flame the top 20 mm before removing the explants/callus and placing them in a sterile Petri dish. It is preferable to accumulate six explants/callus pieces in each dish. Usually after 4 weeks in culture the explants incubated on medium with 2,4-D have formed a substantial callus and each piece should be divided into two or three pieces of not less than 5 mm×5 mm. The cells on the lower side of the explant towards the centre are frequently necrotic in appearance and not suitable for sub-culture. These necrotic areas should be separated and discarded. Ensure that the lid of the Petri dish is replaced after each manipulation. Transfer the developing explant/callus to fresh medium in a culture tube. Flame the top 20 mm of the tube containing fresh medium and discard the aluminium foil cap. Using flamed forceps remove two pieces of callus and place these apart on the surface of the medium towards the end of the tube. Take a square of aluminium foil and close the open end of the culture tube ensuring that a good seal is formed, flame the foil end place the tube in the rack. This procedure should be repeated until all the cultures have been transferred. Prolonged culture of carrot tissue produces large calluses. The procedures described for the first subculture of the developing callus are similar to those used for the routine sub-culture of established callus lines with a transfer period of 4 weeks.

Scheduling

Event	Timing
Isolation of fresh explants from carrot root	Day 0
First subculture	Day 28
Marked proliferation of callus	c. Day 42–48

Fig. 1.3 Items for the sterile transfer room

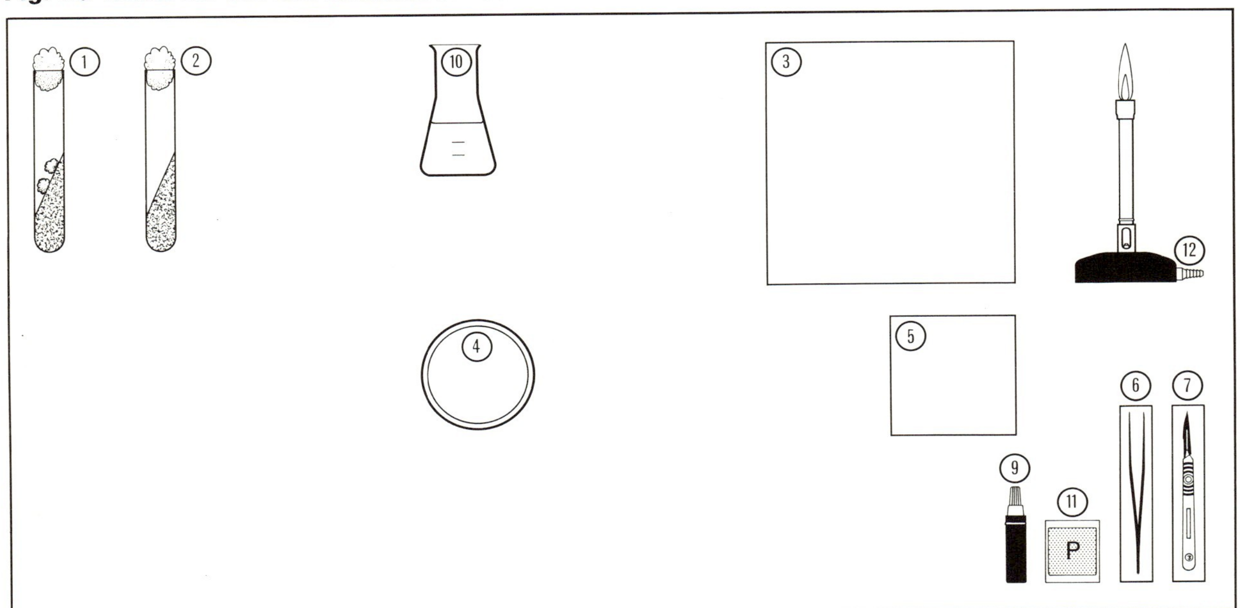

Isolation of established callus	c. Day 91–98
Transfer period of established carrot callus	4 weeks

Recording of Results

Copy down the details of the experiment in a note book, recording the starting date of the experiment, the duration, number of cultures, number of infections, and the different treatments employed. Make visual observations of the cultures at weekly intervals for 4 weeks, recording changes in the morphology of the cultures. Determine the fresh weight of the cultures initially and after 4 weeks of growth.

Questions and Comments

From which regions of the freshly isolated explants does the callus originate?
Do all the explants grow at a similar rate?

References

Gautheret, R.J.: La Culture des Tissues Végétaux. Paris: Masson & Cie 1959

Reinert, J.: Morphogenesis in Plant Tissue Cultures. Endeavour *XXL*, 85–90 (1962)

Yeoman, M.M., Macleod, A.J.: Tissue (Callus) Cultures – Techniques, pp. 31–59. In: Plant Tissue and Cell Culture (Street, H.E. ed.) Botanical Monographs *11*. Oxford: Blackwell 1977

Experiment 2 Growth and Cell Division in Cultured Artichoke Explants

Tissues isolated from intact plants can provide excellent material for the study of a variety of physiological processes. If large amounts of sterile tissue are required then the choice of a donor organ is strictly limited and includes the tap roots of carrot, parsnip and chicory or the stem tubers of potato and Jerusalem artichoke. Tissues removed from the tap roots of chicory, or parsnip, or carrot, respond well in culture but are not clonal in origin and also tend to be heterogeneous with respect to the constituent cellular population. Potato can provide clonal material and while possessing a high degree of cellular uniformity does exhibit a variability of response between tubers. However, certain varieties of Jerusalem artichoke possess substantial tubers from which large amounts of highly uniform sterile tissue can be removed. Such tissue responds readily and uniformly to a variety of nutrients and growth substances and has been used extensively for studies on the effects of plant growth regulators on a range of physiological processes. It is also a simple matter to propagate a population of plants from a single tuber within a few growth seasons and produce large amounts of clonal material.

In addition to the advantageous properties already described, explanted material from dormant Jerusalem artichoke possesses another important property. A significant proportion (60%) of the constituent cells of each explant divide with a high degree of synchrony when placed in contact with a simple nutrient medium containing sucrose and 2,4-D and this has enabled extensive studies to be made of the structural and molecular events which take place during cell division.

Materials and Equipment

Sterile Items

(For sterilisation procedures and composition of medium see Appendix)

1 20 sheets of tissue paper (200 × 200 mm)
2 10 paper towels (c. 370 × 230 mm)
3 10 Petri dishes, 90 mm diameter, either glass or plastic
4 50 sheets of aluminium foil (100 × 100 mm)
5 3 wide necked Erlenmeyer flasks (100 ml) containing 15 ml of liquid medium M_{ART} and a magnetic stirring bar (25 mm) encased in PTFE

Fig. 2.1 Items for the sterile transfer room

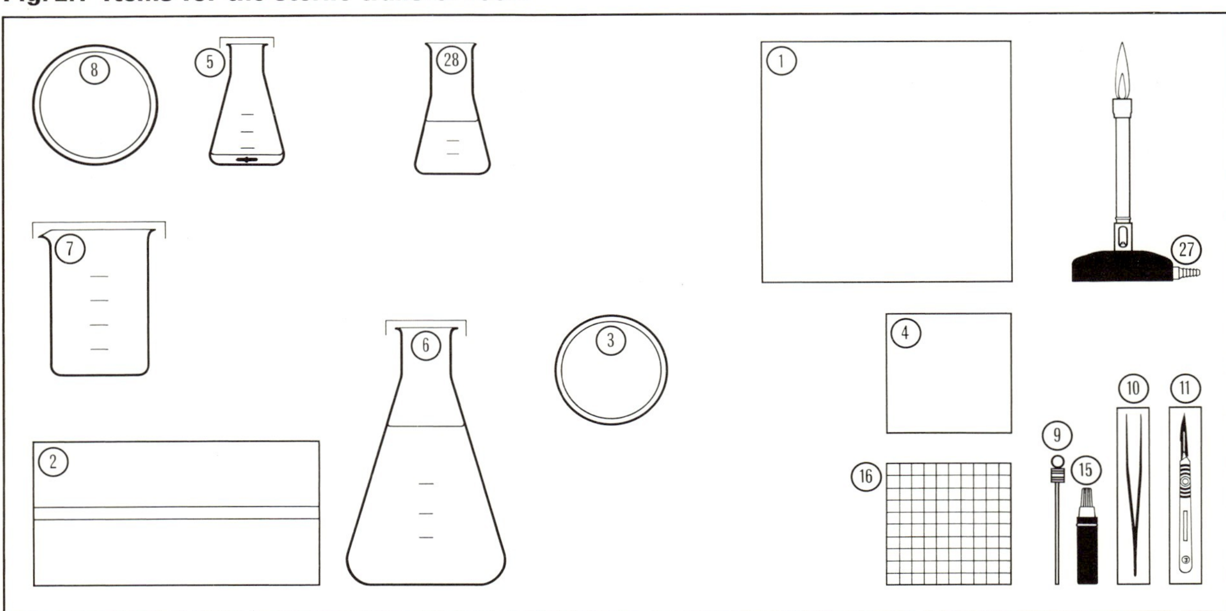

6 3 l of sterile distilled water contained in 3 Erlenmeyer flasks (1,000 ml)
7 1 beaker (1,000 ml)
8 6 crystallising dishes with lids (90 mm in diameter and 50 mm deep)
9 2 cork borers or canulas, 2 mm inside diameter
10 3 pairs of forceps (120–150 mm)
11 3 scalpels (c. 150 mm)

Non-sterile Items

12 Two large tubers (50–100 g) from the Jerusalem artichoke *(Helianthus tuberosus)* bought from the local market or preferably grown in open ground, harvested in early winter (November–December) and stored in barely damp sand in polythene bags, in the dark at 4 °C. A particularly good variety for the experiments is, Large French White, but any strain of Jerusalem artichoke tuber can be used. If the tubers are purchased from the market the risk of contamination is greater and the tubers should be checked for sterility before embarking on large complex experiments
13 3 glass beakers (500 ml)
14 1,000 ml solution of sodium hypochlorite, approximately 20% (v/v); commercial bleach preparations such as Ace or Domestos can be used at a dilution of one part to five of water
15 1 waterproof marking pen
16 1 square of mm graph paper, 100 × 100 mm, covered with a transparent waterproof film
17 3 magnetic stirrers to rotate cultures at c. 250 rpm
18 1 nylon nail brush
19 1 polythene 'washing up' bowl
20 18 glass specimen tubes (50 × 25 mm) with closures
21 1 hot air oven at 95 °C
22 1 analytical balance
23 1 desiccator
24 18 glass or plastic specimen tubes (70 × 15 mm) each containing 2 ml of a 5% (w/v) solution of chromium trioxide in water (chromic acid)
25 1 haemocytometer slide with double coverslips
26 3 Pasteur pipettes complete with rubber teats
27 1 Bunsen or ethanol burner
28 1 Erlenmeyer flask (150 ml) containing 100 ml 95% ethanol

The items for the sterile transfer room are laid out as shown in Fig. 2.1.

Experimental Procedures

Select tubers of uniform shape and similar size discarding any small, diseased, damaged or irregularly shaped individuals. Scrub the selected tubers with a nylon nail brush in running tap water to remove surface soil, wrap in tissue paper and immerse in 20% (v/v) sodium hypochlorite contained in 500 ml beakers for 25–30 min. Transfer the beakers with the tubers to the sterile room during the sterilisation procedure. After 25–30 min immersion remove the tubers, after peeling off the tissue paper, and transfer to a 1,000 ml beaker containing ca. 500 ml of sterile distilled water. Agitate the beaker by hand to ensure that the hypochlorite and its degradation products are removed from the surface of the tubers. Pour off the liquid and replace with a fresh 500 ml aliquot of sterile distilled water, repeat this procedure four times (4 ×). After the final rinse remove the tubers individually, wiping them dry with a sterile paper towel and sterile tissue paper and place each tuber in a sterile crystallising dish covered with a lid.

Fig. 2.2 Isolation of explants

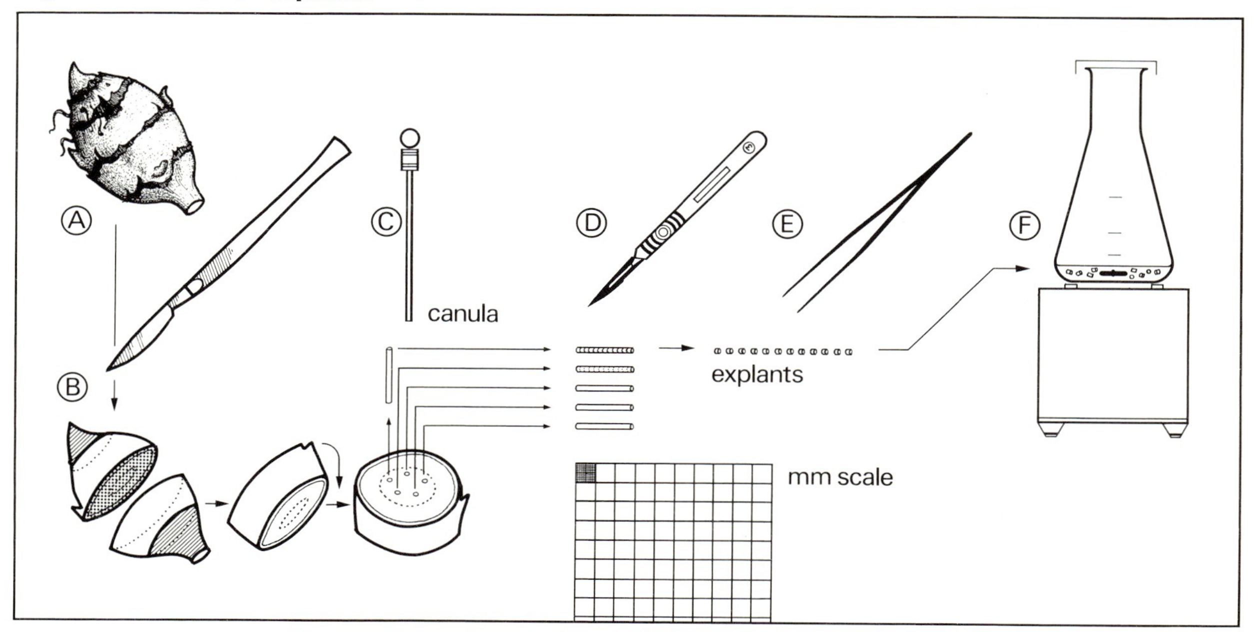

Transfer single tubers (Fig. 2.2A) to a sterile Petri dish, cut off both ends (c. 10 mm) with a scalpel and discard. Divide the tuber transversely into pieces approximately 30 mm in length and remove a series of cylinders 2 mm in diameter from the central region of the tuber keeping at least 2 mm from the cambium (Fig. 2.2B and C). Place the cylinders from each tuber in a fresh sterile Petri dish. Cut up the cylinders into pieces 2.4 mm long using the mm scale (Fig. 2.2D). Transfer 120 explants, using forceps (Fig. 2.2E), to each 100 ml Erlenmeyer flask, flame the neck, seal with a double layer of aluminium foil, flame again and place on top of a magnetic stirrer in the dark at 25 °C±1° (Fig. 2.2F).
A variety of growth parameters may be examined using this culture system over a period of 10 days. Sample 10 explants at the start of the experiment (day 0) and subsequently at 2, 3, 5, 7, and 10 days. Remove 10 explants from each experimental flask using strictly aseptic procedures. Place the explants on two thicknesses of tissue paper and gently roll each individual over using forceps to remove the superficial moisture. Weigh each explant to the nearest tenth of a mg (fresh weight). Place five of the explants in a corked 50 mm×25 mm glass specimen tube of known weight and dry in an oven at 95 °C for 48 h, transfer to a desiccator and allow to cool to room temperature. Reweigh the tube containing the dried explants to the nearest tenth of a mg (dry weight). Place the remaining five explants into a small glass tube containing 5% chromic acid and determine the cell number of the explants according to the technique described in the Appendix.

Scheduling

Event	Timing
Preparation of explants	Day 0
Removal of samples. (10 for fresh weight; 5 of these used to determine dry weight and 5 for cell number)	Day 0, 2, 3, 5 7 and 10

Recording of Results

Copy down the details of the experiment into a notebook, recording the starting date and duration of the experiment, and the number of cultures. Plot your results graphically and determine the rate of increase of fresh weight, dry weight, and cell number per explant. Calculate the average fresh and dry weight per cell and plot these secondary data graphically.

Questions and Comments

Does the rate of increase in fresh weight remain constant during the culture period?
Are the increases in fresh weight and cell number, and dry weight and cell number directly related?
Did you discover any evidence of cellular differentiation in the cultures?

References

Aitchison, P.A., Macleod, A.J., Yeoman, M.M.: Growth patterns in tissue (callus) cultures, pp. 267–306. In: Plant Tissue and Cell Culture (Street, H.E. ed.). Botanical Monographs *11*. Oxford: Blackwell 1977
Yeoman, M.M.: Early development in callus cultures. Int. Rev. Cytol. *29*, 383–409 (1970)
Yeoman, M.M., Evans, P.K.: Growth and differentiation of plant tissue cultures. II. Synchronous cell division in developing callus cultures. Ann. Bot. *31*, 323–332 (1967)
Yeoman, M.M., Dyer, A.F., Robertson, A.I.: Growth and differentiation of plant tissue cultures. I. Changes accompanying the growth of explants from *Helianthus tuberosus* tubers. Ann. Bot. *29*, 265–276 (1965)

Experiment 3 Initiation and Establishment of Cell Suspension Cultures of Carrot *(Daucus carota)*

Experiment 3 Initiation and Establishment of Cell Suspension Cultures of Carrot *(Daucus carota)*

The culture of plant tissues in an agitated liquid medium eliminates many of the disadvantages ascribed to the culture of tissues on agar. Movement of the tissue in relation to the nutrient medium facilitates gaseous exchange, removes any polarity of the tissue due to gravity, and eliminates nutrient gradients within the medium and at the surface of the cells. The incubation of a friable callus in a liquid nutrient medium agitated on a shaking machine, will eventually give a suspension of cells which are more amenable to experimental manipulation than callus grown on agar. The technique described here is one in which a callus of *Daucus carota* is suspended in a liquid nutrient medium and is subjected to a series of manipulations to ensure the production of an actively growing cell suspension consisting of single cells and very small aggregates of 2–15 cells.

Materials and Equipment

Sterile Items

(For sterilisation procedures and composition of medium see Appendix)

1 5 established friable *Daucus carota* callus cultures 7–10 days after subculture on $M_{DAUC}+5\times10^{-8}$ g/ml 2,4-D, supported on a rack at 25°
2 5 wide necked Erlenmeyer flasks (250 ml) (at each sub-culture) containing 60 ml of M_{DAUC} medium$+5\times10^{-8}$ g/ml 2,4-D
3 5 measuring cylinders (100 ml) with foil caps
4 2 spoon spatulas which are at least 100 mm longer than the measuring cylinders
5 25 sheets of aluminium foil (100×100 mm)
6 2 pieces of nylon, or stainless steel sieves with a porosity of 250 μm, which can be inserted into the opening of the measuring cylinders
7 5 Petri dishes (90 mm in diameter)

Non-Sterile Items

8 1 waterproof marking pen
9 1 rotary shaking machine (speed of rotation 80–100 rpm)
10 1 Bunsen or ethanol burner
11 1 Erlenmeyer flask (150 ml) containing 100 ml 95% ethanol
12 1 roll of parafilm

Fig. 3.1 Items for the sterile transfer room

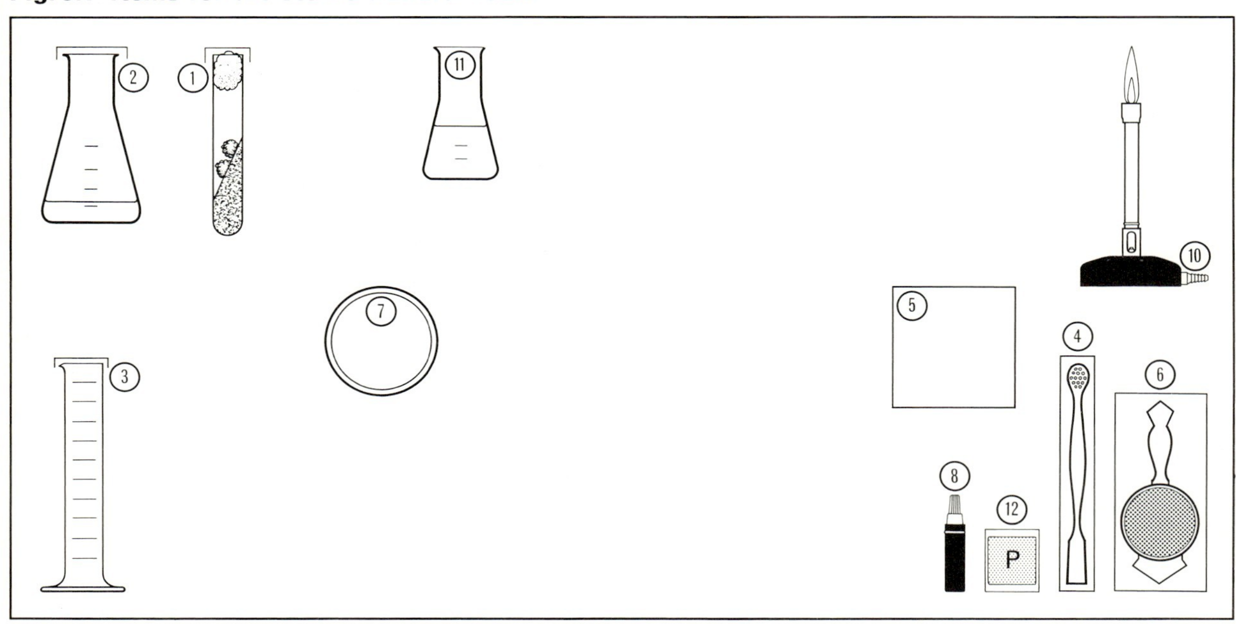

The items for the sterile transfer room are laid out as shown in Fig. 3.1.

Experimental Procedures

Remove the culture tubes one at a time from the rack and flame the top 20 mm before emptying the callus into a sterile Petri dish. Repeat this operation until the cultures have been removed from all five tubes and placed separately in Petri dishes. Take a 250 ml conical flask containing 60 ml of M_{DAUC} medium, +2,4-D, flame the neck, remove the foil cap, flame again and transfer 3–4 pieces of callus (c. 1 g each) from one Petri dish using the spoon spatula. Flame the neck, close the top of the flask with a new piece of foil ensuring that a good seal is made and flame again, finally secure the foil cap with parafilm.

Repeat this operation until all five flasks with the liquid medium have been inoculated and then place the flasks on the rotary shaking machine in the dark or in low intensity light at 25 °C. After 7 days transfer the flasks from the shaker to the sterile room. Remove the parafilm, flame and the remove the foil cap from each flask. After flaming the neck of the flask again, pour the contents of each culture flask through the sieve into a sterile 100 ml measuring cylinder and replace the foil cap. Ensure that the sieve remains sterile when it is put onto the measuring cylinder by replacing the foil cap.

Allow the contents to settle for c. 10 min and then pour off the supernatant. Empty the residue of cells in the measuring cylinder into a new 250 ml flask containing 60 ml of fresh medium, flame the neck and recap with foil as before. Repeat this operation with all five flasks and replace them on the shaker. At the next subculture (after 7 days) repeat the procedure with the measuring cylinder but take only one fifth (1/5th) of the residual cells as the inoculum.

As a rule, after 3–4 transfer periods of 7 days under the conditions described, the carrot suspensions contain only single cells and small cell aggregates. These cultures are suitable for the determination of growth rates using a series of parameters (see Appendix).

The growth pattern of suspension cultures depends upon the cell density per ml of nutrient at inoculation. Therefore it is important to determine the effect of inoculum size (cell number) on subsequent growth of the culture. With a sub-optimal inoculum density growth is either slow or does not occur as all, while supra-optimal amounts of inocula shorten the lag phase but the growth rate of these cultures is diminished and growth stops earlier. For carrot cell suspension a range between 10^5 and 3×10^5 cells/ml or 100 and 300 mg (fresh weight) of inoculum in a volume of 60 ml of fresh medium $M_{DAUC} + 5 \times 10^{-8}$ g/ml, 2,4-D and a transfer period of 7 days is appropriate.

After 3–4 culture periods, carrot cell suspensions can simply be subcultured by the transfer of 10 ml of cell suspension in sterile, wide-bored 10 ml pipettes, to 50 ml of fresh medium.

Scheduling

Event	Timing
Initiation of cell suspensions and determination of initial total cell number and of packed cell volume	Day 0
Determination of total cell number and of packed cell volume	Days 1, 2, 4 and 7
First transfer of cultures	Day 8
Fourth transfer of cultures	Day 29

Recording of Results

Copy down the details of the experiment in a notebook recording the starting date and the duration of the experiment, number of cultures and the different treatments employed. Make visual observations of the suspension cultures at weekly intervals for 3 weeks, recording changes in the morphology and the number of cultures infected. Determine the packed cell volume (see Appendix) at the beginning and at the end of the each transfer period.

After 3 subcultures, estimate the total cell number initially and at days 1, 2, 4, and 7, and plot these values against time. Examine the relationship between the density of the inoculum and the subsequent pattern of growth.

Questions and Comments

Explain the pattern of growth obtained in experiments in which the inoculum density is varied.

What type of curve would result from daily counts of the total cell number?

What changes in the growth curve (cell number and packed cell volume) will occur if the transfer period is prolonged to 14 days?

References

Henshaw, G.G., Iha, K.K., Mehta, A.R., Shakeshaft, D.J., Street, H.E.: Studies on the growth in culture of plant cells. I. Growth patterns in batch propagated suspension cultures. J. Exp. Bot. *17*, 362–377 (1966)

Street, H.E.: Cell (Suspension) Culture-Techniques, pp. 61–102, In: Plant Tissue and Cell Culture (Street, H-E., ed.) Botanical Monographs *11*, Oxford: Blackwell 1977

Torrey, J.G., Reinert, J.: Suspension culture of higher plant cells in synthetic media. Plant Physiol. *36*, 483–491 (1961)

Experiment 4 Isolation and Culture of Single Cells and Examination of the 'Conditioning' Effect

Single isolated cells of higher plants, like those of bacteria or fungi, provide excellent tools for the study of cloning and genetic stability, the nature of the interactions between the host cell and the pathogen, as well as the more exacting growth requirements of single cells. In the previous experiment it was clearly demonstrated that there is a minimal size of inoculum for the initiation and growth of cell suspension cultures. This suggests that besides the requirement for a suitable nutrient medium, growth also depends upon interactions between the cells in culture. Although the exact nature of these interactions cannot yet be defined, it is clear that the nutrient medium is changed by the presence of callus or suspended cells and this constitutes a 'conditioning' process which encourages the growth of single cells and small aggregates.
Single cells can be isolated and cultured using different methods. In the following experiment, cells of a friable callus culture of *Daucus carota* are plated on agar. This technique permits the direct microscopical observation of the growing cells.

Materials and Equipment
Sterile Items
(For sterilisation procedures and composition of medium see Appendix)

1 5 culture tubes (150×25 mm) each containing 2 callus cultures of *Daucus carota* on M_{DAUC}+ 5×10^{-8} g/ml 2,4-D, 10 days after the last transfer
2 9 plastic Petri dishes (55 mm in diameter) containing c. 10 ml agar (1%) M_{DAUC} medium+5×10^{-8} g/ml 2,4-D. This thickness of the agar layer will allow the observation of the cells by the objectives with the lowest magnification of the inverted microscope
3 1 plastic Petri dish (55 mm in diameter) *for conditioned medium, wrapped in aluminium foil*
4 2 Erlenmeyer flasks (25 ml) containing 10 ml liquid M_{DAUC} medium+5×10^{-8} g/ml 2,4-D
5 1 beaker (600 ml) containing 250 ml water for warming the conditioned medium
6 4 Pasteur pipettes with teats and the end (4 mm) below the teats filled with cotton wool

Fig. 4.1 Items for the sterile transfer room

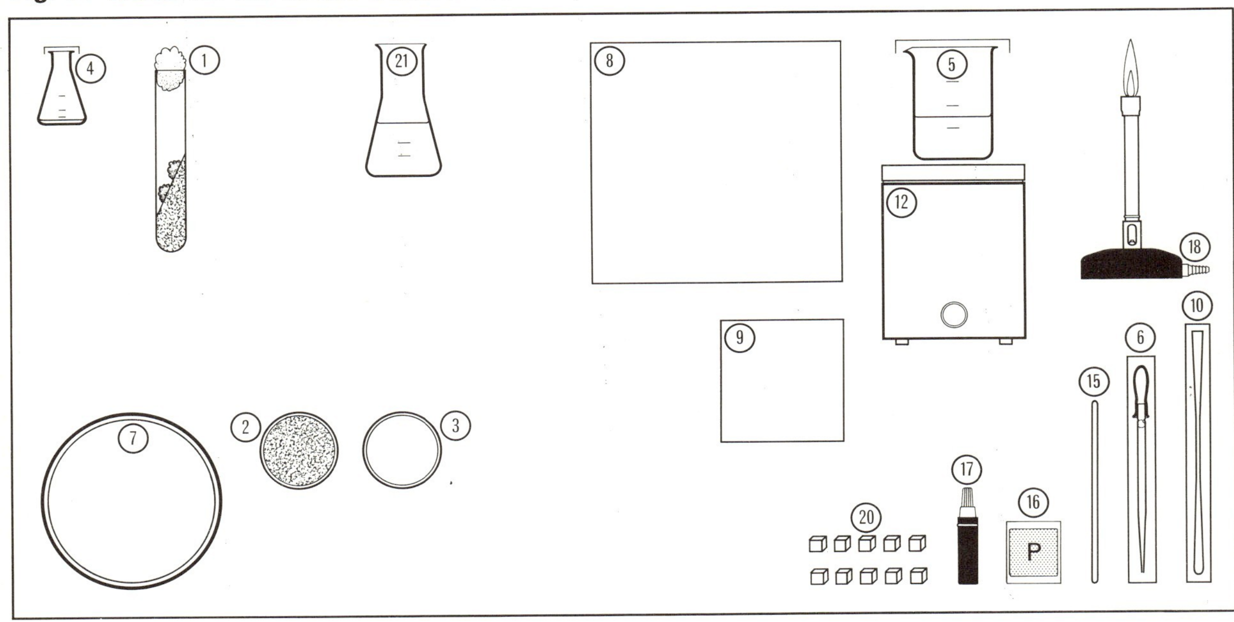

7 3 Petri dishes 140 mm in diameter
8 20 sheets of tissue paper (200×200 mm)
9 50 sheets of aluminium foil (100×100 mm)
10 2 thin and flexible spatulas preferably stainless steel (200×3 mm)

Non-Sterile Items

11 2 large Petri dishes (c. 140 mm in diameter)
12 1 electrically heated 'hotplate'
13 1 inverted microscope with a co-ordinate system
14 1 analytical balance for the determination of weights between 10 mg and 250 mg
15 1 glass rod (200×2 mm)
16 1 roll of parafilm
17 1 waterproof marking pen
18 1 Bunsen or ethanol burner
19 1 Erlenmeyer flask (250 ml) containing 150 ml 95% ethanol
20 10 small boxes (10×10×10 mm) folded from aluminium foil so that the opening is at one of the sides. These are sterilised by dipping in ethanol, flamed and kept in a large sterile Petri dish. (For the sterile weighing procedure)
21 1 Erlenmeyer flask (150 ml) containing 100 ml 95% ethanol

The items for the sterile transfer room are laid out as shown in Fig. 4.1.

Experimental Procedures

Use of a Conditioned Medium

Take one of the culture tubes containing 2 carrot cultures on M_{DAUC} medium+2,4-D, 10 days after the last transfer, and carefully remove all cells and debris off the agar with a spatula. Cover the tube again with aluminium foil and place in a boiling water bath. Immediately the agar has melted take the culture tube out of the boiling water, remove the foil cap, flame the open end thoroughly and pour the contents of the tube (10 ml) into a sterile Petri dish (55 mm) and allow the medium to resolidify. Set out 3 Petri dishes (55 mm), containing fresh M_{DAUC} medium+2,4-D together with the single dish with the 'conditioned' medium. Using a Pasteur pipette, place 2 drops of liquid medium into the middle of each of the 4 Petri dishes. Weigh out portions of the callus (2×10 mg, 1×50 mg, 1×200 mg) using the sterile procedure and retain within the aluminium boxes. Then transfer the callus pieces onto the drops of liquid medium in the Petri dishes. Mix the callus and the liquid medium carefully with a spatula or glass rod to ensure that the cells and cell aggregates in suspension can be distributed evenly. Spread out the suspension over the agar plates with the glass rod, firstly in the form of a star (Fig. 4.2A) and then by distributing the cells over the whole surface (Fig. 4.2B and C). Take care not to damage the agar surface! Seal all Petri dishes with parafilm.

Use of a 'Nurse' Culture

Repeat the procedure of plating callus (10 mg) in 2 drops of M_{DAUC} medium+2,4-D on an agar surface (M_{DAUC}+2,4-D) in a Petri dish (55 mm) but in addition, transfer 4 pieces of callus (50 mg) and position on the agar surface as shown in Fig. 4.3. After inoculation seal each of the 5 Petri dishes immediately with parafilm, place within larger Petri dishes (140 mm) and keep in the dark at 28 °C for a period of 10 days. Check the growth of the isolated single cells daily using the following procedure. Firstly, using an inverted microscope to ensure the easy location of single cells on the agar layer draw a circle around each single cell to be

Fig. 4.2 Spreading suspension on agar plate

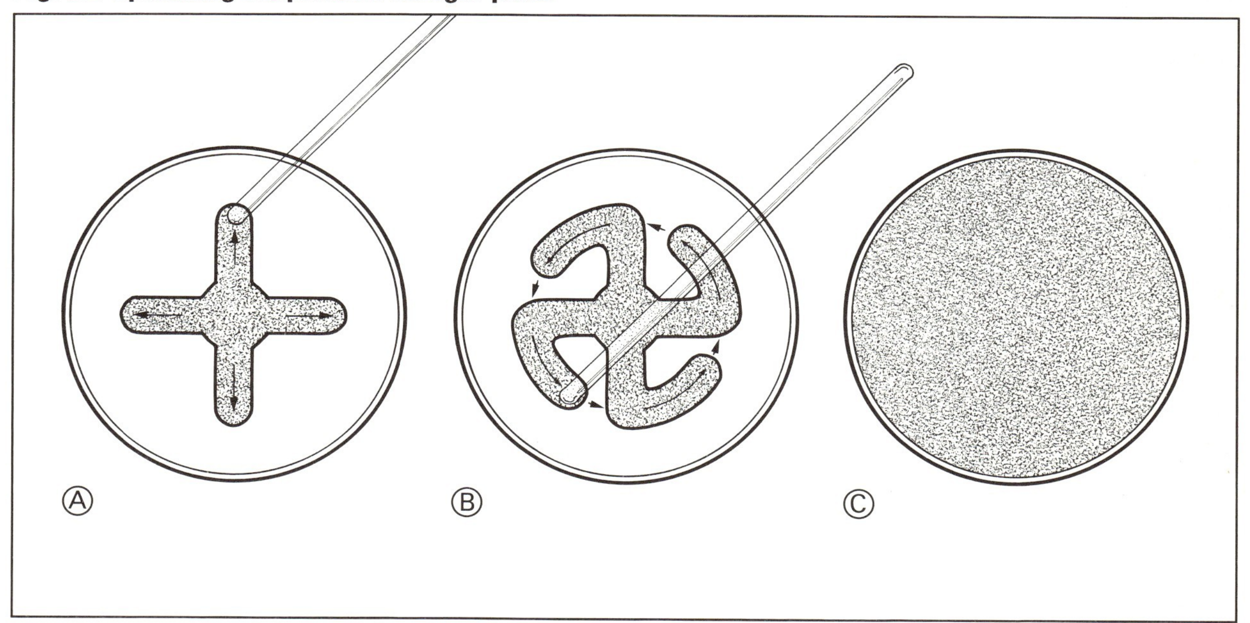

observed on the lower side of the Petri dish with a fine waterproof marking pen. Secondly insert the pen through the opening of the microscope stage from below and draw around the circumference of the opening; make sure that the dish does not move during drawing. Now, place a co-ordinate system (120×120 mm) sub-

Fig. 4.3 Position of callus pieces on agar

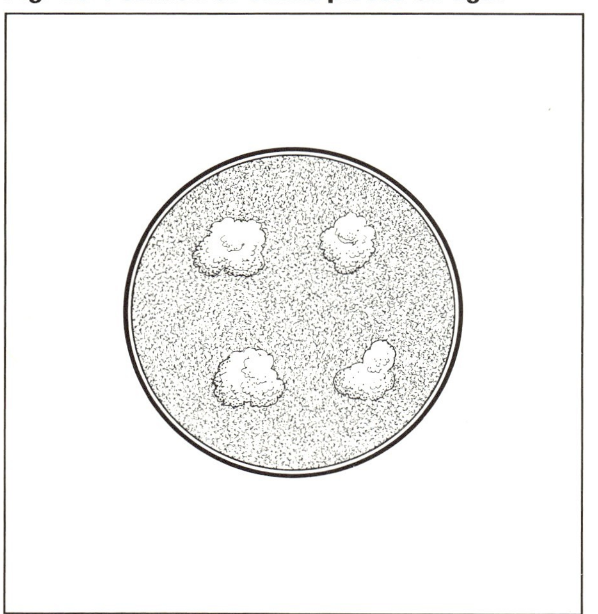

Fig. 4.4 Microscope with coordinate system

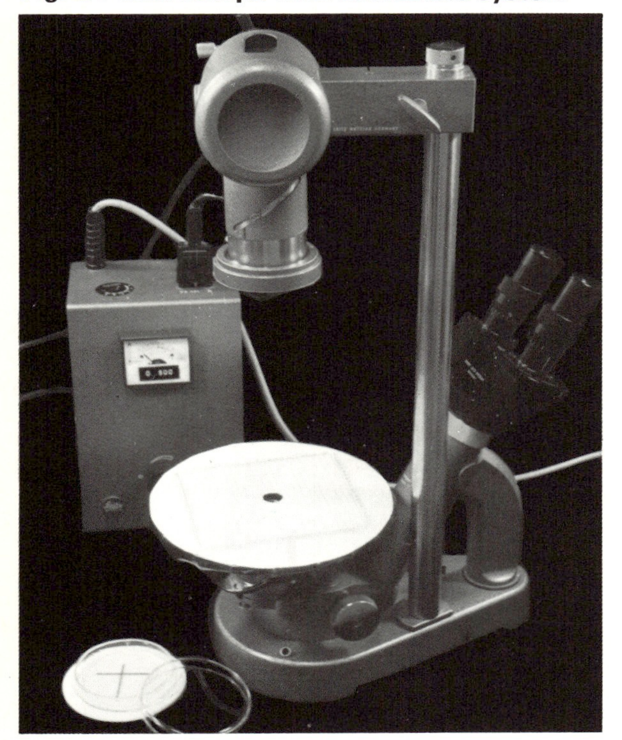

divided into 5 mm squares with its centre (ca. 15 mm) directly above the hole in the microsope stage (Fig. 4.4). Because of the thickness of the agar-layer only an objective with a low magnification can be used. This arrangement in combination with a 'T-marker' (place the Petri dish with the cell cultures on top of a box or table with right angled corners and draw a T-shape on the lower surface of the dish with a fine waterproof marking pen) drawn on the lower side of the Petri dish containing the cell cultures can be used to quickly localise single cells for observation (Fig. 4.5). Do not turn the Petri dish upside down at any time! On the first day after inoculation select and mark at least 12 undamaged single cells for observation, some of these cells will die during the first 2 days after isolation. New cells may be located and marked after day 3 to restore the number of observed cells to 12. Cells will start to divide between day 3 and 5 of culture.
Growing single cells change shape and it is sometimes difficult to recognise these cells after a few divisions, therefore, in addition to the procedure of marking by circles and the co-ordinate system, make drawings of each of the isolated cells with their surroundings, beginning at day 3 and subsequently at daily intervals during the remainder of the culture period of 10 days.

Fig. 4.5 Localisation of single cell in culture

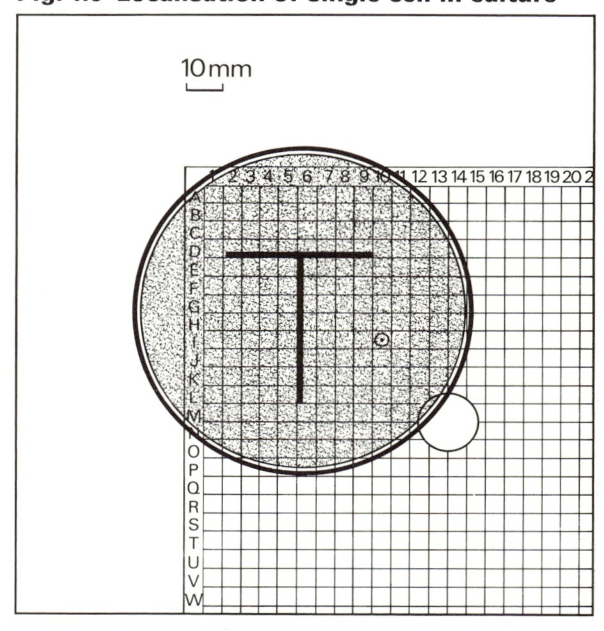

Scheduling

Event	Timing
Initiation of cultures, marking of 12 isolated cells	Day 0
Beginning of daily observations, replacement of dead cells by new selection	Day 3
End of daily observations	Day 10

Recording of Results

Copy down the details of the experiment in a note-book recording the starting date, the duration, the number of cultures, the different treatments employed, and the number of contaminations. Record changes in the morphology, of each cell and its daughters. Plot the number of cell divisions a) against culture time and b) against the weight of the inoculum.

Questions and Comments

The optimal culture time for conditioned 'medium' is 10 days. What is the reason for this restriction?
What changes may occur in the medium during the 'conditioning' process? Discuss.

References

Backs-Hüsemann, D., Reinert, J.: Embryobildung durch isolierte Einzelzellen aus Gewebekulturen von *Daucus carota.* Protoplasma *70,* 49–60 (1970)

Hildebrandt, A. C.: Single Cell Culture, Protoplasts and Plant Viruses, pp. 581–597. In: Applied and Fundamental Aspects of Plant Cell, Tissue, and Organ Culture. (Reinert, J., Bajaj, Y. P. S., eds.). Berlin-Heidelberg-New York: Springer 1977

Reinert, J.: Growth of single cells from *Haplopappus gracilis* and *Vitis vinifera* tissues on synthetic media. International Conference on Plant Tissue Cultures. Pennsylv. State Univ. 1963. (White, P. R., Grove, A. R., eds.). Berkeley: McCutchan 1965, pp. 61–67

Torrey, J. G.: Cell division in isolated single cells in vitro. Proc. Natl. Acad. Sci. USA *43,* 887–891 (1957)

III Bioassay Systems for Cytokinins

Cytokinins are phytohormones which induce and maintain cell division in tissue cultures in the presence of auxin. The importance of this class of substances in regulating growth and development was discovered from work with the cultured pith of the stem of tobacco *Nicotiana tabacum.* Most natural cytokinins are 6-amino-purine derivatives. In addition a number of 6-amino-purine derivatives which do not occur in plants also act as cytokinins, e.g., kinetin and 6-benzyl amino purine.

Formulae of cytokinins:

Main ring:

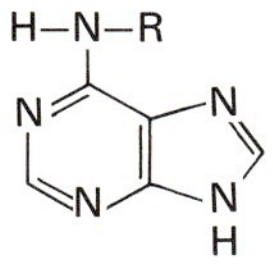

H–N–R

N N N N H

Natural compounds:

1) $R = -CH_2-CH=C(CH_3)(CH_2OH)$

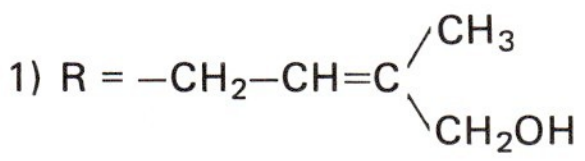

6–(4–Hydroxy–3–methyl–trans–2–butenylamino) purine (Zeatin)

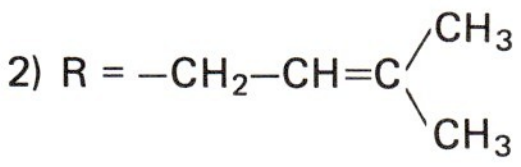

2) $R = -CH_2-CH=C(CH_3)(CH_3)$

6–(8–methyl–2–butenylamino) purine (Dimethylallyladenine)

Synthetic compounds:

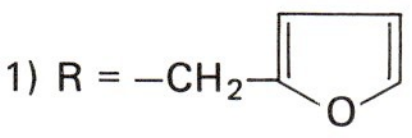

1) $R = -CH_2-$ (furan ring, O)

6–Furfurylaminopurine (Kinetin)

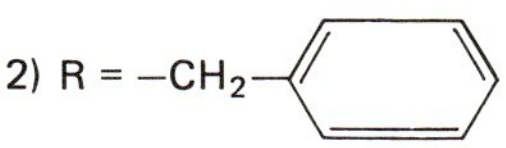

2) $R = -CH_2-$ (benzene ring)

6–Benzylaminopurine (Benzyladenine)

However, it is also important to remember that some synthetic 6-amino-purine derivatives do not act as cytokinins, e.g-. 6-dimethyl-amino-purine and 2-dimethyl-amino-6-hydroxy purine. Cytokinins are not only concerned with cell division but are also involved in the control of organ formation and the prevention of senescence in leaves. They are also involved in the breaking of dormancy in buds. The quantitative determination of cytokinins, which are present in very low concentrations in plants, is usually made using bioassay procedures which are based on the stimulation of cell division or other effects by these hormones. Three of these methods, which are incidentally the most successful, involve tissue culture procedures. These are based on the stimulation of cell division by cytokinins, the first uses freshly isolated tobacco pith explants, the second established callus derived from tobacco pith and the third mature callus from soybean cotyledons.

Tobacco pith (without vascular traces), tobacco callus and soybean callus will not grow on basic culture media with auxin alone and require, in addition, a supply of a cytokinin. Under strictly controlled conditions and within certain limits of concentration the increase in fresh weight of the tobacco pith or callus, or soybean callus, is directly proportional to the $\log_{10}$ of the cytokinin concentration in the medium and this provides the basis for the bioassay procedures. Two bioassays will be described, the first utilising established callus of soybean and the second tobacco callus.

Experiment 5 Soybean Bioassay System

Materials and Equipment

Sterile Items

(For sterilisation procedures and composition of medium see Appendix).

1 20 pieces (5 cultures) of established soybean callus (c. 10×10×10 mm) three weeks after the last transfer, growing on 40 ml of agar medium $M_{SOY}+2\times10^{-6}$ g/ml NAA+5×10^{-7} g/ml kinetin in 100 ml Erlenmeyer flasks. In order to raise the sensitivity of the test the volume of the medium may be reduced to not less than 25 ml
2 20 Erlenmeyer flasks (100 ml) containing 40 ml of M_{SOY} medium+2×10^{-6} g/ml 2,4-D solidified with 1% agar. The range of concentrations of kinetin in the assay media is 0–2×10^{-6} g/ml (Table 5.1). It is important to use an established line of cultures isolated from the cotyledons of *Glycine max* (Var. Acme). For the isolation procedure of new callus lines, see Miller (1963)
3 5 Petri dishes, 90 mm in diameter, either glass or plastic
4 50 sheets of aluminium foil (100×100 mm)
5 2 stainless steel spatulas (160 mm)
6 1 Petri dish 140 mm in diameter

Table 5.1

Treatment No.	No. of Flasks	Concentration of Kinetin (g/ml)	(μg/l)
1	4	0	(0)
2	4	2×10^{-9}	(2)
3	4	2×10^{-8}	(20)
4	4	2×10^{-7}	(200)
5	4	2×10^{-6}	(2,000)

Non-Sterile Items

7 3 trays to hold 20 wide-necked Erlenmeyer flasks (100 ml)
8 1 waterproof marking pen
9 1 analytical balance for determining the fresh weight of the culture

Fig. 5.1 Items for the sterile transfer room

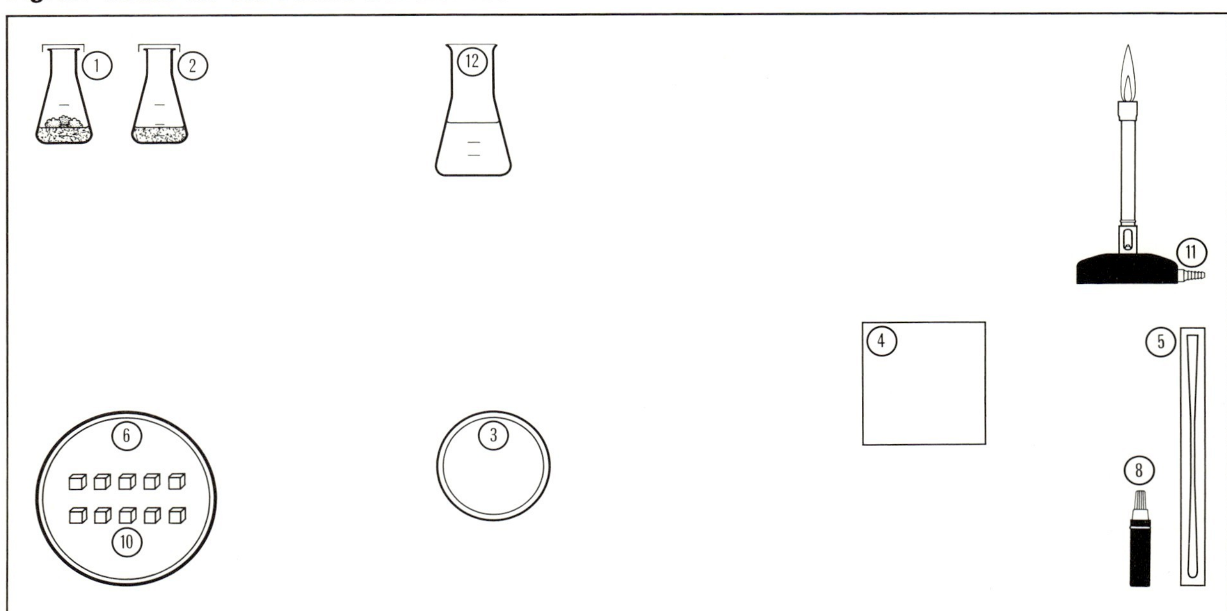

10 10 small boxes (10×10×10 mm) folded from aluminium foil so that the opening is at one of the sides. These are sterilised by dipping in ethanol, flamed and kept in a large sterile Petri dish
11 1 Bunsen or ethanol burner
12 1 Erlenmeyer flask (150 ml) containing 100 ml 95% ethanol

The items for the sterile transfer room are laid out as shown in Fig. 5.1.

Experimental Procedures

Only use callus cultures which are between 3 and 4 weeks old (since the last subculture) and which have begun to turn brown; in these cultures the kinetin content of the callus will be approaching zero and therefore kinetin will not be carried over with the subculture. Flame the neck of the Erlenmeyer flask, discard the aluminium foil cap and transfer the callus from the surface of the medium to a sterile Petri dish with a spatula. Separate the peripheral parts, divide these pieces of callus into four equal fragments of approximately 10 mg (2×2×2 mm) and immediately weigh each piece to the nearest mg under sterile conditions. It is very important at this stage to randomise the population of callus fragments (80) to eliminate differences between the 20 pieces of established callus. This procedure requires care and skill in order to avoid contamination. Another method may be used which is less accurate. Here the relationship between the weight and size of several pieces of callus is determined under non-sterile conditions and this relationship is subsequently used as a pattern for preparation of callus fragments under conditions of strict asepsis. Transfer four pieces of callus to the surface of the medium of an experimental flask, flame the neck, seal with aluminium foil and flame again. Ensure that weight of the fragments is marked on the outside of the flask. Incubate all of the flasks at 25 °C, 1,600 lux, in a 12 h day for 21 days. At the end of the culture period determine the fresh weight of each piece of callus to the nearest mg and calculate the average values for each treatment.

Scheduling

Event	Timing
Transfer callus pieces of known weight to the assay media	Day 0 (3 weeks after last subculture)
Harvest the cultures and determine the fresh weight of individual pieces	Day 21
Calculate and process the data	Day 22

Recording of Results

Copy the details of the experiment into a notebook recording the starting date and the duration of the experiment, number of cultures and the different treatments employed. Insert the individual 21 day fresh weight values (X_1) and the means ($\bar{X}$) for each experimental treatment into Table 5.2. Plot the mean values ($\bar{X}$) on a graph in which fresh weight is the ordinate and kinetin concentration the abscissa. In order to determine whether one treatment is significantly different from the others, statistical analyses must be employed. Complete Table 5.2 and calculate the standard error for each treatment using the formula provided. If possible consult a statistician and discuss the experiment with him before proceeding.

Table 5.2

Kinetin (g/ml)	X_1	$\bar{X}$	SM
0 (Control)			
2×10^{-9}			
2×10^{-8}			
2×10^{-7}			
2×10^{-6}			

S^2 = variance
SM = standard deviation of the mean standard error)
X_1, X_2, X_3, X_4 = individual values
$\bar{X}$ = mean value
n = number of values

$$S^2 = \frac{\Sigma(X^2) - \frac{\Sigma x^2}{n}}{n-1}$$

$$M = \sqrt{\frac{S^2}{n}}$$

Questions and Comments

What would be the effect of using cultures less than 3 weeks old?

Explain why extending the concentration range of kinetin would lead to a breakdown in linearity.

Make a list of the possible sources of error in this bioassay procedure. How could you improve the bioassay techniques?

References

See under Experiment 6.

Experiment 6 Tobacco Callus Bioassay System

Materials and Equipment

Sterile Items

(For sterilisation procedures and composition of medium see Appendix)

1 3 cultures of an established callus of tobacco (*Nicotiana tabacum* Wisconsin No. 38) isolated originally from stem pith and three weeks since the last subculture, growing on 100 ml of M_{TOB} agar medium+2×10^{-6} g/ml IAA+2×10^{-7} g/ml kinetin in Erlenmeyer flasks (250 ml). (Murashige and Skoog, 1962)

2 24 Erlenmeyer flasks (250 ml) containing 100 ml of M_{TOB} medium+2×10^{-6} g/ml IAA with kinetin added. The range of concentrations of kinetin in the assay media is 0–6×10^{-8} g/ml (Table 6.1)

3 10 Petri dishes 90 mm in diameter, either plastic or glass

4 50 sheets of aluminium foil (100×100 mm)

5 2 stainless steel spatulas (160 mm)

6 1 Petri dish, 140 mm in diameter

Table 6.1

Treat-ment No.	No. of Flasks	Concentration of Kinetin	
		g/ml	µg/l
1	3	0	(0)
2	3	4×10^{-9}	(4)
3	3	6×10^{-9}	(6)
4	3	1×10^{-8}	(10)
5	3	1.5×10^{-8}	(15)
6	3	2×10^{-8}	(20)
7	3	3×10^{-8}	(30)
8	3	6×10^{-8}	(60)

Non-Sterile Items

7 3 trays to hold 24 Erlenmeyer flasks (250 ml)

8 1 analytical balance for determining the fresh weight of the cultures

9 1 waterproof marking pen

10 1 Bunsen or ethanol burner

Fig. 6.1 Items for the sterile transfer room

11 1 Erlenmeyer flask (150 ml) containing 100 ml 95% ethanol
12 10 small boxes (10×10×10 mm) folded from aluminium foil so that the opening is at one of the sides. These are sterilised by dipping in ethanol, flamed and kept in a large sterile Petri dish (140 mm)

Under these experimental conditions there is a linear relationship between $\log_{10}$ of the cytokinin concentration (5–25 μg) and callus fresh weight.
The items for the sterile transfer room are laid out as shown in Fig. 6.1.

Experimental Procedures
Use only the upper pale coloured parts of callus cultures which are between 3 and 4 weeks old. In these cultures the kinetin content of the callus, especially those parts furthest away from the medium will be approaching zero and therefore it is unlikely that kinetin will be transferred to the assay cultures. Flame the neck of each Erlenmeyer flask containing the tobacco callus, discard the aluminium foil cap and transfer the callus from the surface of the medium to a sterile Petri dish with a spatula. Remove fragments of callus of approximately 20–30 mg from the top part of the culture and weigh each piece immediately to the nearest mg under sterile conditions (as in Experiment 5). Transfer each piece to the surface of the medium of an experimental flask, flame the neck, seal with aluminium foil and flame again. Ensure that the weight of each fragment is marked on the outside of the flask. Incubate all of the cultures in the dark at 25 °C for 4 weeks. At the end of the culture period determine the fresh weight of each piece of callus to the nearest mg and calculate the means for each treatment.

Scheduling

Event	Timing
Transfer pieces of callus of known weight to assay media	Day 0 (3 weeks after last subculture)
Harvest the cultures and determine the fresh weight of individual pieces	Day 28
Calculate and process results	Day 29

Recording of Results
Using the tobacco callus data repeat the procedure and statistical analysis set out in Experiment 5.

Questions and Comments
See under Experiment 5.

References

Fox, J. E.: The Cytokinins, pp. 85–123. In: Physiology of Plant Growth and Development. (Wilkins, M. B. ed.). London: McGraw-Hill 1969

Letham, D. S.: Chemistry and Physiology of Kinetin-like Compounds. Annu. Rev. Plant Physiol. *18*, 349–364 (1967)

Miller, C. O.: Kinetin and Kinetin-like Compounds. In: Modern Methods of Plant Analysis, Vol. *6*, 194–202. (Linskens, H. F., Tracey, M. V. eds.). Berlin-Heidelberg-New York: Springer 1963, pp. 194–202

Murashige, T., Skoog, F.: A revised medium for rapid growth and bioassays with tobacco tissue cultures. Physiol. Plant *15*, 473–497 (1962)

Patau, K., Das, N. K., Skoog, F.: Induction of DNA synthesis by kinetin and indoleacetic acid in excised tobacco pith tissue. Physiol. Plant *10*, 949–966 (1957)

Skoog, F., Miller, C. O.: Chemical regulation of growth and organ formation in plant tissues cultured in vitro, pp. 118–131. In: Biological Action of Growth Substances. (Porter, H. K. ed.). S.E.B. Symp. *11*, Cambridge: CUP 1957

IV Morphogenesis in Vitro: Studies on Regeneration

IV Morphogenesis in Vitro: Studies on Regeneration

Experiment 7 Embryogenesis in Cultured Cells of Carrot

Somatic cells of tissue cultures from freshly isolated carrot explants, and also from established callus strains, can be induced to form adventive embryos by manipulation of the medium. The pathway of development of these somatic embryos closely follows that taken by the fertilised egg cell in the embryo sac of the developing carrot seed, showing typical globular and torpedo stages. This phenomenon provides excellent proof for the totipotency of higher plant cells. Since this first demonstration of embryogenesis in carrot cultures there have been many examples of embryogenesis in cell cultures of other higher plants and the number of reports is still increasing.

Embryogenesis in vitro has been investigated using different nutrient media and it has been clearly shown that in carrot and some other species two of the constituents of the medium are critical in determining the response; the auxin and the source of nitrogen, provided M_{DAUC} and M_{HAP} media are used (see Appendix). Obviously in this case it is the balance between these compounds which determines whether embryos will be formed or not. Roots may also be formed in addition to embryos. The morphogenetic capability of cells in culture usually changes during prolonged cultivation. In carrot cells, for instance, it can disappear completely 30–40 weeks after the isolation of the explant but can be re-induced by manipulation of the medium; thus indicating that in this example the genetic information for embryogenesis is always present and that it is the epigenetic factors which are essential for the regulation of embryogenesis in carrot cultures. In this experiment the effects of two medium constituents on embryogenesis in established callus lines will be examined.

Materials and Equipment

Sterile Items

(For sterilisation procedures and composition of media see Appendix)

1 Carrot cultures which have lost the ability to form embryos during prolonged cultivation (c. 1 year), 10 growing on M_{DAUC} medium $+5\times10^{-8}$ g/ml 2,4-D and 14 on M_{HAP} medium $+5\times10^{-8}$ g/ml 2,4-D

Fig. 7.1 Items for the sterile transfer room

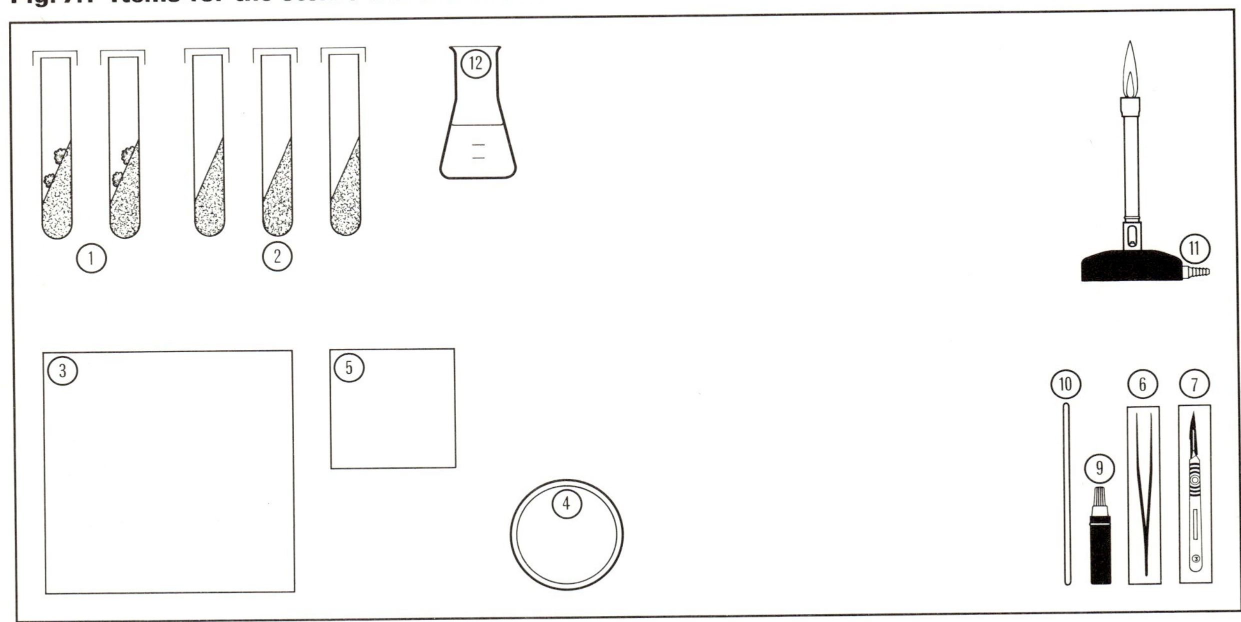

2 Culture tubes (150×25 mm) containing 10 ml of agar medium sloped at an angle of 25°

No. of Tubes	Medium
20	M_{DAUC} +2,4-D
20	M_{DAUC} −2,4-D
10	M_{HAP} +2,4-D

3 25 sheets of tissue paper (200×200 mm)
4 5 Petri dishes, 90 mm in diameter
5 50 sheets of aluminium foil (100×100 mm)
6 2 pairs of forceps (120–150 mm)
7 3 scalpels (c. 250 mm)

Non-Sterile Items

8 5 racks, preferably plastic or metal, each to hold 12 culture tubes (150×25 mm) at an angle of approximately 25°
9 1 waterproof marking pen
10 2 glass rods (150×5 mm)
11 1 Bunsen or ethanol burner
12 1 Erlenmeyer flask (150 ml) containing 100 ml 95% ethanol
13 1 Erlenmeyer flask (250 ml) containing 150 ml of freshly prepared liquid M_{DAUC} medium to check embryo stages

The items for the sterile transfer room are laid out as shown in Fig. 7.1.

Experimental Procedure

Transfer the callus cultures using the technique described in Exp. 1. Each callus will provide 4–6 pieces (approximately 5×5×5 mm) taken from the edge of the culture. Make sure that only the pale, proliferating callus is transferred. These cells can easily be separated from the culture with a scalpel or by gently pushing with forceps, and collected in a sterile Petri dish. Transfer two of these pieces to each agar slope and discard the rest of the callus. Accumulate these pieces of callus in sterile Petri dishes. Ensure that the lids of the Petri dishes are replaced immediately after each manipulation and that the aluminium foil is tightly wrapped around the open end of each culture tube and flame the neck of the tube thoroughly. Repeat the process until all the cultures on M_{DAUC}+2,4-D and M_{HAP}+2,4-D medium have been transferred (Fig. 7.2).

At intervals of one month using aseptic procedures, remove small samples of callus from cultures growing on M_{DAUC} medium+2,4-D, M_{DAUC} medium without 2,4-D and M_{HAP} medium+2,4-D (control), close the culture tubes immediately after this manipulation. Place these samples individually into Petri dishes containing 10 ml of liquid M_{DAUC} medium and stir gently with a glass rod until the cells float apart. This treatment separates the different stages of embryo development. Observe the suspension of cells with a binocular microscope or an inverted microscope.

If 'old' callus which has lost the ability to undergo embryogenesis in vitro is not available, 'younger' cultures (12 weeks after the isolation of the carrot explant) can be used. However, these react differently and usually form embryos on M_{DAUC}+2,4-D, but not on M_{HAP}+2,4-D medium. Therefore only the treatments in part B of Fig. 7.2 will provide clear cut results.

Fig. 7.2

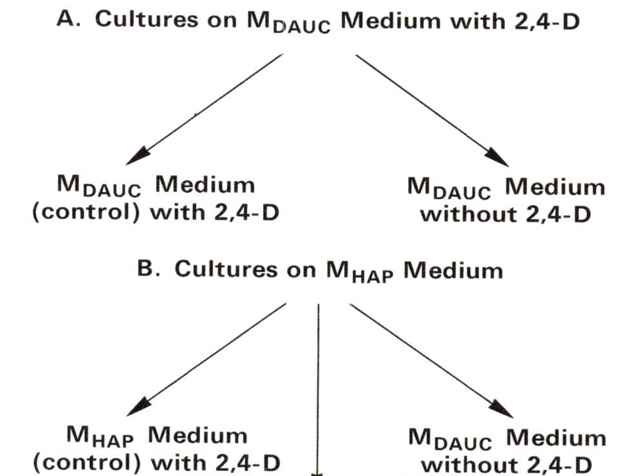

Scheduling

Event	Timing
Transfer of cultures to M_{DAUC}+2,4-D, M_{HAP}+2,4-D (control) and M_{DAUC} without 2,4-D	Day 0
Sample the cultures and record incidence of embryogenesis	Every 28 days
Emergence of globular embryo stages	c. Day 62–68
Formation of mature embryos	c. Day 124–130

Recording of Results

Copy down the details of the experiment into a notebook recording starting date, the number of cultures, the duration of the experiment and the treatments employed. Make microscopical observations of the cultures at 28 day intervals and make drawings of the different stages of embryo development. Record the frequency of contamination.

Questions and Comments

What other factor(s) besides auxin determines embryo formation in these carrot cultures?

Discuss why only some of the cells of the carrot cultures form embryos in vitro.

Can you detect any signs of embryogenesis in the control cultures?

Carrot is a member of the Umbelliferae which forms genetically heterogeneous populations and no pure lines. This provides a source of variation within the population of carrots and the cell cultures derived from them which is expressed in differences in growth rate and in embryogenic potential of the cultures.
Cultures with fully developed embryos are able to grow into mature green plants after transfer to M_{DAUC} medium without 2,4-D (solidified with agar) and kept in white light at 25 °C.

References

Halperin, W., Wetherell, W. F.: Adventive embryony in tissue cultures of the wild carrot *(Daucus carota)*. Am. J. Bot. *51*, 274–283 (1964)

Narajanaswamy, S.: Regeneration of plants from tissue cultures, pp. 179–206. In: Applied and Fundamental Aspects of Plant Cell, Tissue and Organ Culture (Reinert, J., Bajaj, Y. P. S., eds.). Berlin-Heidelberg-New York: Springer 1977

Reinert, J.: Untersuchungen über die Morphogenese an Gewebekulturen. Ber. Dtsch. Bot. Ges. *71*, 65 (1958). Mitgliederversammlung, 15.

Reinert, J.: Über die Kontrolle der Morphogenese und die Induktion von Adventivembryonen an Gewebekulturen aus Karotten. Planta *53*, 318–333 (1959)

Reinert, J., Backs-Hüsemann, D., Zerban, H.: Determination of embryo and root formation in tissue cultures from *Daucus carota*, pp. 261–268. In: Les Cultures de Tissues des Plantes. Paris: Colloques Internationaux du C.N.R.S., *193* (1971)

Reinert, J., Bajaj, Y. P S., Zbell, B.: Aspects of Organisation–Organogenesis, Embryogenesis, Cyto-differentiation, pp. 389–427. In: Plant Tissue and Cell Culture (Street, H. E. ed.). Botanical Monographs *11*. Oxford: Blackwell 1977

Steward, F. C., Mapes, M. O., Mears, K.: Growth and Development of Cultured Cells. II. Organisation in cultures grown from freely suspended cells. Am. J. Bot. *45*, 705–708 (1958)

Experiment 8 The Tobacco Pith System

Experiment 8 The Tobacco Pith System

Many plant cell cultures are able to regenerate roots, shoots, embryos and complete plantlets. It is also known that the morphogenetic capabilities of plant cells can be regulated by changes to the levels of growth regulatory substances in the nutrient media on which the tissues are cultured. Classical examples of systems which can be manipulated to regenerate are explants and established calluses from tobacco and carrot. The first demonstration that the balance of two growth regulatory substances in the medium could determine the nature of the subsequent regeneration was made with tobacco callus. In this system it appears to be a delicate balance between the levels of IAA and a cytokinin which determines whether roots, shoots or only callus will be produced. The question of whether this is a fundamental property of all higher plants or a particular effect characteristic of only a few species remains unresolved. The material used here, tobacco pith parenchyma, is particularly suitable for class use because of the uniformity of response between experiments.

Materials and Equipment

Sterile Items

(For sterilisation procedures and composition of medium see Appendix)

1 6×10 rimless culture tubes (150×25 mm) containing 10 ml of 0.8% agar medium M_{DAUC} of the following composition (see Table 8.1, A–F)

Table 8.1

Treatment	No. of culture tubes	Kinetin mg/l	IAA mg/l
A	10	0	0
B	10	0	0.2
C	10	0.2	0
D	10	0.2	3.0
E	10	3.0	0.2
F	10	3.0	0.02

Fig. 8.1 Items for the sterile transfer room

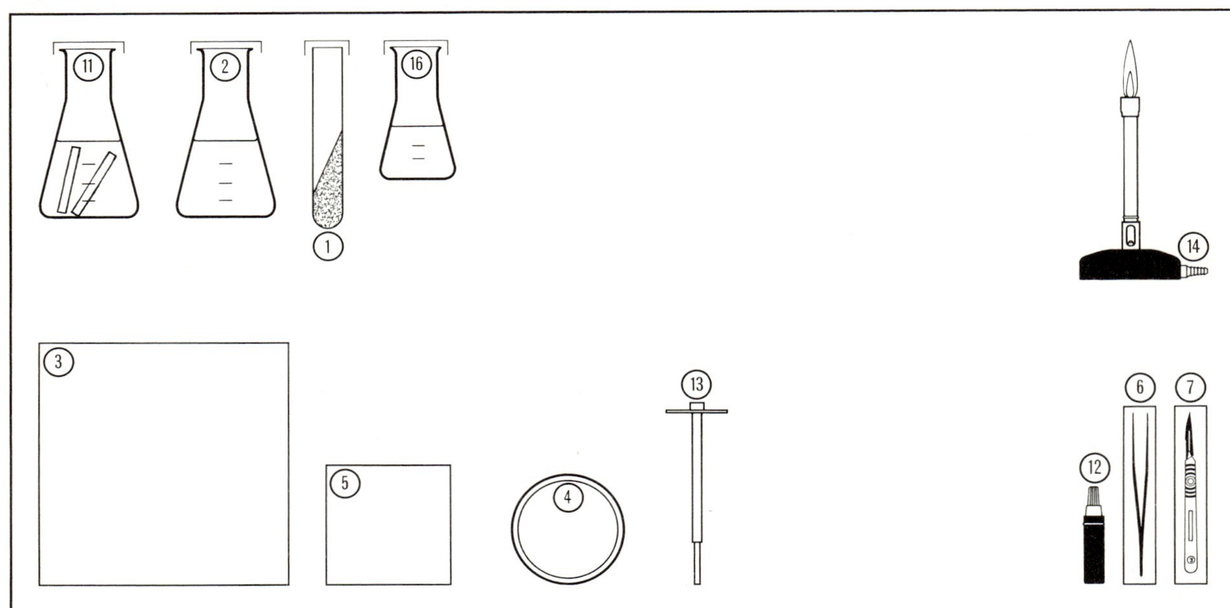

2 4 wide-necked Erlenmeyer flasks (250 ml) containing 200 ml of water
3 40 sheets of tissue paper (200×200 mm)
4 10 Petri dishes, 90 mm in diameter, either glass or plastic
5 60 sheets of aluminium foil (100×100 mm)
6 2 pairs of forceps (120–150 mm)
7 2 scalpels (150 mm)

Non-Sterile Items

8 1 vigorously growing plant of *Nicotiana tabacum*, Var. 'Badischer Burley', or 'White Burley', 8–10 weeks old and approximately 1 m high
9 3 racks, preferably plastic or metal, to hold 12 culture tubes (150×25 mm) at an angle of approximately 25°
10 1 wide necked Erlenmeyer flask (250 ml) containing 200 ml of 70% ethanol
11 1 Erlenmeyer flask (250 ml) containing 200 ml of a solution of sodium hypochlorite, 20% (v/v). Commercial bleach preparations such as Ace or Domestos can be used at a dilution of 1 part to 5 of water
12 1 waterproof marking pen
13 1 Corkborer, size to suit diameter of tobacco pith
14 1 Bunsen or ethanol burner
15 1 metal scalpel (150 mm)
16 1 Erlenmeyer flask (150 ml) containing 100 ml of 95% ethanol

The items for the sterile transfer room are laid out as shown in Fig. 8.1.

Experimental Procedures

Cut the terminal 300 mm of the main stem from the tobacco plant. Separate and discard the leaves, side buds and the apical 100 mm with the meristem, and dip the cut ends of the explant in molten wax to seal off. Submerge the remaining stem piece of 200 mm for 20 s in 70% ethanol and then transfer to the solution of sodium hypochlorite. Transfer the tobacco stem in the hypochlorite to the sterile room and mark each culture tube with the designated treatment and the date. Remove the piece of stem from the hypochlorite solution after 30 min and rinse twice with sterilised water, finally drying thoroughly with the tissue paper. With a sharp scalpel cut off and discard 10 mm from each end of the stem and sub-divide the remaining part into segments, 20 mm in length. While holding the stem segment with a pair of sterile forceps, remove a cylinder within the pith parenchyma using the corkborer.

Transfer these cylinders to a sterile 90 mm Petri dish and cut into discs, 2–3 mm thick, with a sharp scalpel. Repeat this procedure until sufficient explants have been accumulated for 10 tubes (2 per culture tube). Always replace the lid of the Petri dish after each manipulation. Remove the closure from a culture tube and flame the open end. While holding the tube at an angle of 45°, using forceps, transfer two explants onto the surface of the agar medium taking care that the basal side ('root' end) is in contact with the agar surface. Seal each tube immediately with a square of aluminium foil which has been flamed before and after it has been placed on the tube. Incubate the cultures in continuous light (2,000 lux) at 25 °C. At the completion of the experimental procedure 120 explants will have been transferred to the 60 tubes containing the different media. The first signs of callus formation will appear on the upper side of the explant a few days after the initiation of the cultures. The callus is clearly visible after 15 days and can be sub-divided after 28 days. Subculture the callus as described in Experiment 1 to fresh media of the same composition. Roots, shoots and plantlets will appear later, after approximately 35–50 days.

Scheduling

Event	Timing
Initiation of cultures	Day 0
Swelling of the explant	c. Day 3–5
Beginning of the proliferation	c. Day 7
Callusing from the upper end	c. Day 10–15
Transfer of the callus	c. Day 28–35
Root formation	c. Day 35–50
Shoot formation	c. Day 42–50*
Plantlet formation (only rarely)	c. Day 42–50*

* Depends on the auxin/cytokinin ratio.

Recording

Copy down the details of the experiment in a notebook, recording the starting date and duration of the experiment, number of cultures, and the different treatments employed. Make visual observations of the cultures at 3–4 day intervals after the first transfer, recording changes in morphology. Make drawings of the cultures at the end of the experiment.

Questions and Comments

What is the overall effect of auxin and kinetin on the growth and morphogenesis of the explants?
Do the roots and/or shoots originate directly from the explant or from the callus?
Do the regenerated shoots form roots?

References

Galston, A.W., Davies, P.J.: Control Mechanisms in Plant Development. Prentice-Hall: Englewood Cliffs, N.J. 1970

Letham, D.S.: Cytokinins and their relation to other phytohormones. BioScience *19*, 309–316 (1969)

Miller, C.O., Skoog, F., von Saltza, M.N., Strong, F.M.: Kinetin, a cell division factor from deoxyribonucleic acid. J. Am. Chem. Soc. *77*, 1392 (1955)

Skoog, F., Miller, C.O.: Chemical regulation of growth and organ formation in plant tissues cultured in vitro. Symp. Soc. Exp. Biol. *11*, 118–131 (1957)

Experiment 9 Culture of Anthers from *Nicotiana tabacum* and the Establishment of Haploid Plants from Embryos

Anthers may be removed aseptically from sterilised flower buds and placed in culture. A proportion of pollen within these anthers, particularly those from Solanaceous species, grow and proceed through a series of developmental stages, eventually giving rise to haploid embryos. This phenomenon has been designated as 'androgenesis'. The culture conditions and manipulations required to produce embryos from anthers in culture are well established for tobacco, *Nicotiana tabacum.* The developmental stage of the pollen within the anthers is critical for androgenesis and it is recognised that an early stage of pollen development is best. For *Nicotiana tabacum* this is the point at which the microspore nucleus divides to give the generative and vegetative nuclei. The embryo usually develops from the vegetative nucleus and proceeds by a series of mitotic divisions. The whole process is also affected by the physiological age of the donor plant because it would seem that the first flowers that are formed possess a higher androgenetic potential than those that develop later.

Conditions for the successful culture of anthers are also critical. The optimal temperature for androgenesis in tobacco pollen is 25°–28 °C and light is not required for the induction of embryos. However, once the embryos have been induced light is necessary for their development. Phytohormones such as auxins and cytokinins, are not stimulatory and indeed may have a marked inhibitory effect for tobacco pollen, while glutamine, arginine and asparagine further androgenesis. However, auxins are required to promote androgenesis in anthers from *Solanum tuberosum* and *Capsicum annuum.* Conditions for the induction and development of androgenic embryos from the pollen vary between species and may even differ between varieties within a species, for example with *N. tabacum* var. Xanthi, light prevents the establishment of plantlets unlike other varieties where light may be stimulatory.

Materials and Equipment

Sterile Items

(For sterilisation procedures and composition of medium see Appendix)

1 25 Petri dishes (50 mm) containing 5 ml of agar

Fig. 9.1 Items for the sterile transfer room

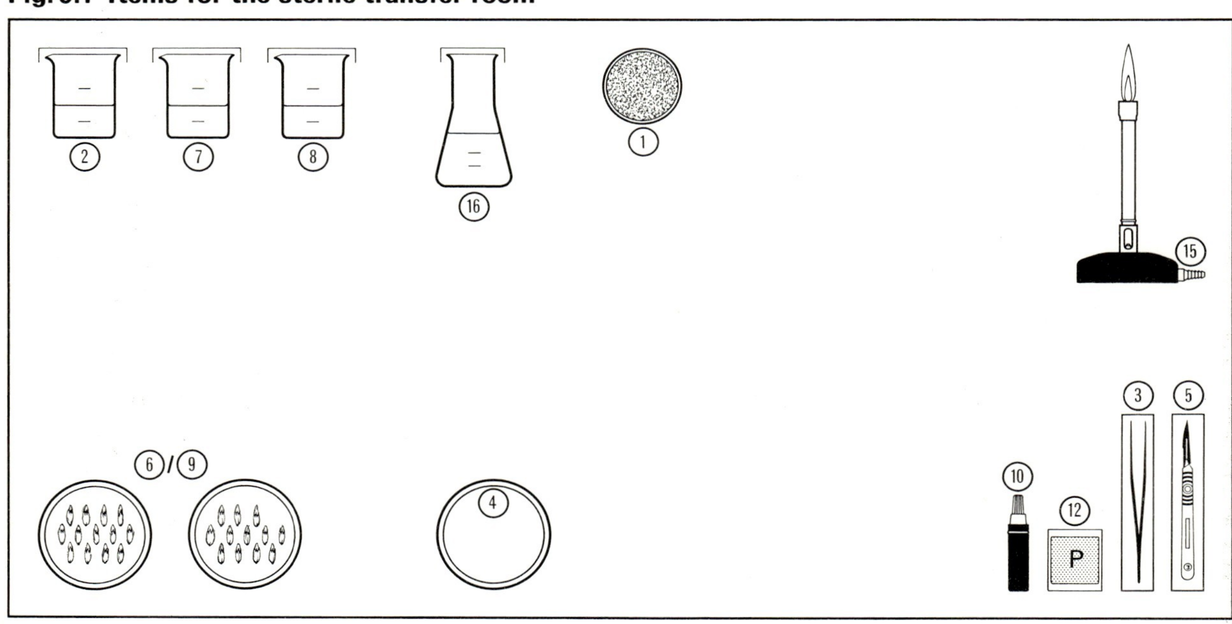

culture medium M_{ANTH}+0.8% Difco Nobel agar

2 3 glass beakers (100 ml) each containing 50 ml distilled water

3 3 pairs of forceps (120–150 mm)

4 5 Petri dishes 90 mm in diameter, glass or plastic

5 3 scalpels (150 mm)

Non-Sterile Items

6 25 closed flower buds (17–22 mm) of *Nicotiana tabacum* collected at the onset of flowering (normally 3–4 months old), when the length of the sepals equals that of the petals (Fig. 9.2)

7 1 glass beaker (100 ml) containing 50 ml 70% ethanol

8 1 glass beaker (100 ml) containing 50 ml 2% sodium hypochlorite solution (v/v)

9 2 Petri dishes 90 mm in diameter, glass or plastic

10 1 waterproof marking pen

11 1 cm ruler

12 1 roll of parafilm

13 6 small plastic pots (50 mm in diameter) filled with sterile potting compost

14 6 glass beakers (100 mm)

15 1 Bunsen or ethanol burner

16 1 Erlenmeyer flask (150 ml) containing 100 ml 95% ethanol

The items for the sterile transfer room are laid out as shown in Fig. 9.1.

Experimental Procedures

Do not remove the flower buds from the plants until the laboratory is ready. Collect the flower buds in a non-sterile Petri dish and measure the length of each individual with a cm ruler. Reject all flower buds which are beginning to open (Fig. 9.2A).

Transfer the Petri dish with the flower buds to the sterile transfer room. Each flower contains 5 anthers and these are normally sterile in closed buds. Sterilise the buds (5 at a time) by immersion in 70% ethanol for 10 s followed immediately by 10 min in 2% (v/v) hypochlorite, then wash by immersion for 5 min in three changes of sterile distilled water (contained in 100 ml beakers). Finally, transfer the buds to a sterile Petri dish. To remove the anthers slit the side of the bud with a sharp scalpel and remove them (Fig. 9.2B). With a pair of forceps place the five anthers with the filaments removed immediately onto the agar medium contained in the Petri dish and close the lid (Fig. 9.2C and D). Seal with parafilm and mark the Petri dish. (Repeat this process until all the material has been used up.) After 3–4 weeks embryos and plantlets will appear from the cultured anthers. At this stage transfer the cultures to a room with a light intensity of c. 2,000 lux with a 14 h day at 28 °C. When approximately 50 mm tall, free the plants from agar by gently washing with running tap water and then transfer them to small pots containing autoclaved potting compost. Cover each plantlet with a small glass beaker (100 ml) to prevent desiccation and maintain in a well-lit humid greenhouse. After one week remove the glass beakers and transfer to larger pots where the plant will mature and eventually flower.

Scheduling

Event	Timing
Initiation of the cultures	Day 0
Emergence of embryos and plantlets	c. Day 21–28
Establishment of plantlets and transfer to pots	c. Day 42–56

Fig. 9.2 Isolation of anthers

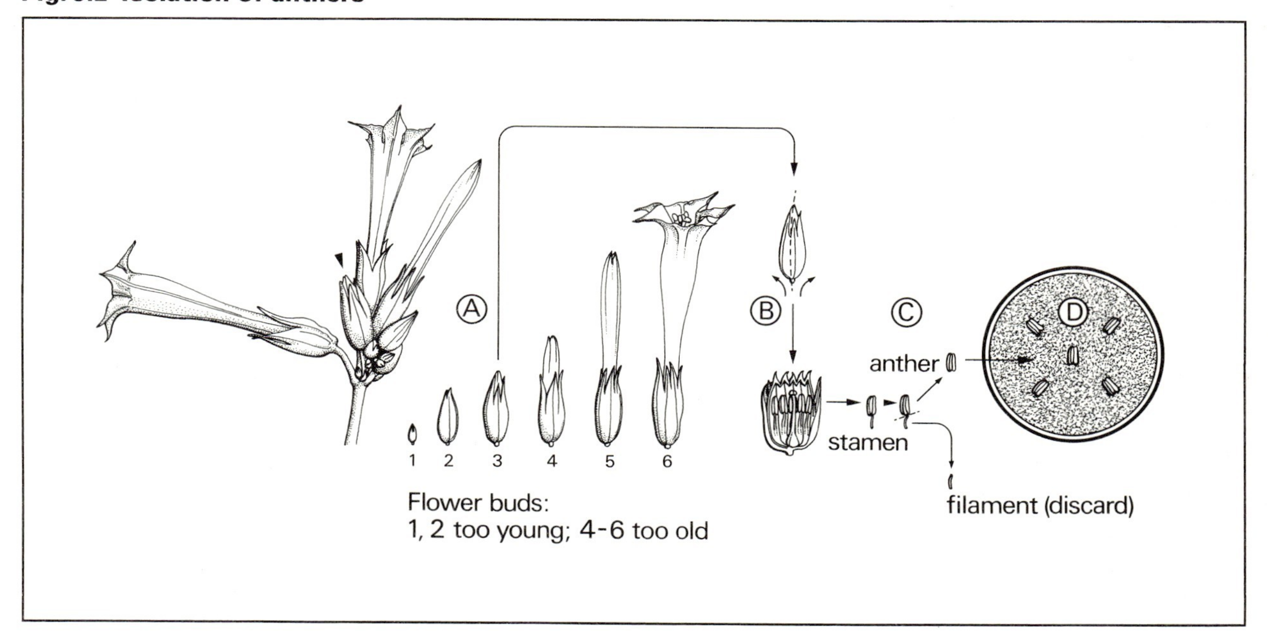

Flowering of haploid plant. c. Day 85–114

Recording of Results

Copy down the details of the experiment into a notebook recording the starting date and duration of the experiment, and the number of contaminated cultures. Make visual observations of the cultures at weekly intervals recording any changes which take place. Record the number of plantlets which arise from each anther and calculate the number of plantlets per dish.

Questions and Comments

Calculate the proportion of pollen grains in each anther which develop into plantlets.
How can you determine whether the plants are haploid or diploid?

References

Hu Han: Advances in Anther Culture Investigations in China, pp. 3–10. In: Proceedings of Symposium on Plant Tissue Culture. Peking, China: Science Press 1978

Nitsch, J. P.: Haploid plants from pollen. Z. Pflanzenzuecht. *67*, 3–18 (1972)

Reinert, J., Bajaj, Y. P. S.: Anther culture: haploid production and its significance. In: Applied and Fundamental Aspects of Plant Cell, Tissue and Organ Culture (Reinert, J., Bajaj, Y. P. S. eds.). Berlin-Heidelberg-New York: Springer 1977, pp. 251–267

Sunderland, N., Wicks, F. M.: Embryoid formation in pollen grains of *Nicotiana tabacum*. J. Exp. Bot. *22*, 213–226 (1971)

Wernicke, W., Kohlenbach, H. W.: Investigations on liquid culture medium as a means of anther culture in *Nicotiana*. Z. Pflanzenphysiol. *79*, 189–198 (1976)

Experiment 10 Vegetative Propagation of Orchids *(Cymbidium)*

Experiment 10 Vegetative Propagation of Orchids *(Cymbidium)*

It is an established fact that the cultivation of explants from monocotyledonous plants is more difficult than from dicotyledonous. This is certainly true for most orchids, although it is possible to excise and culture successfully parts of individuals from a few genera including the genus *Cymbidium*. It has also been shown that explanted floral meristems and protocorms can be cultured on a simple medium containing B vitamins. It is thought that the B vitamins replace the nutrient contribution from the symbiotic fungus which normally grows in close association with the plant. Cultures of several *Cymbidium* species have been initiated from the main bud and 'side meristems' or buds of the so-called 'pseudo bulbs' which develop at the base of the shoot of adult plants. During culture these 'meristems' pass through a developmental sequence similar to that of an orchid embryo during seed germination. A protocorm stage can be clearly distinguished and this is followed by the development of a plantlet. Protocorms, and to a lesser degree very young shoots of *Cymbidium*, are excellently suited for the conventional techniques of vegetative propagation because they possess a high potential for regeneration. If these structures are sub-divided, each piece will re-iterate the development of the seed, with a protocorm stage followed by shoot formation. Although propagation by conventional vegetative means is more rapid than from seed, the yield of clonal material is low; however, the techniques of micro-propagation, using protocorms or meristems, can produce large amounts of genetically homogeneous hybrid material fairly rapidly. An added advantage of using meristems is that the resulting plants are usually virus free.

Materials and Equipment

Sterile Items

(For sterilisation procedures and composition of medium see Appendix)

1 2 rimless culture tubes (150×25 mm) containing 10 ml of agar medium M_{CYM} sloped at an angle of 25°
2 20 sheets of tissue paper (200×200 mm)
3 3 Erlenmeyer flasks (250 ml) capped with aluminium foil, each containing 200 ml distilled water

Fig. 10.1 Items for the sterile transfer room

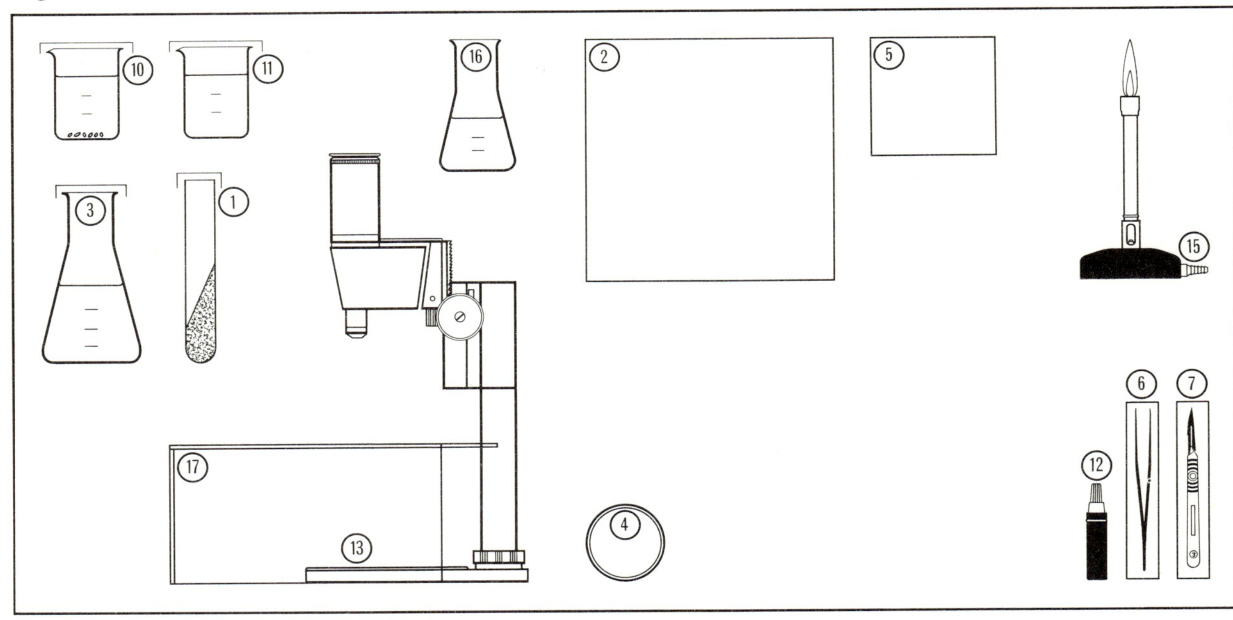

4 6 Petri dishes 50 mm in diameter
5 10 sheets of aluminium foil (100×100 mm)
6 2 pairs of forceps (120–150 mm)
7 2 scalpels (150 mm)

Non-Sterile Items

8 1 bulb of *Cymbidium* (for each person or group). This material may be purchased from most horticultural suppliers or orchid growers and will remain viable, for this experiment, for at least 10 days. Most varieties will respond to the culture procedures employed in this experiment
9 1 rack, preferably plastic or metal, to hold 2 rimless culture tubes (150×25 mm) at an angle of approximately 25°
10 1 glass beaker (250 ml) containing 200 ml of a 20% solution (v/v) of a commercial bleach preparation such as Ace or Domestos. Add one drop of Tween 20 or similar wetting agent to the aqueous solution
11 1 glass beaker (250 ml) containing 200 ml of 95% ethanol
12 1 waterproof marking pen
13 1 binocular dissecting microscope
14 1 metal scalpel (150 mm)
15 1 Bunsen or ethanol burner
16 1 Erlenmeyer flask (150 ml) containing 100 ml 95% ethanol

The items for the sterile transfer room are laid out as shown in Fig. 10.1.

In addition to the above items a perspex (Plexiglas) sterile shield (Fig. 10.2 and item 17 in Fig. 10.1) is also required to facilitate the dissection of the material under aseptic conditions using a binocular microscope.

Fig. 10.2 Microscope with sterile shield

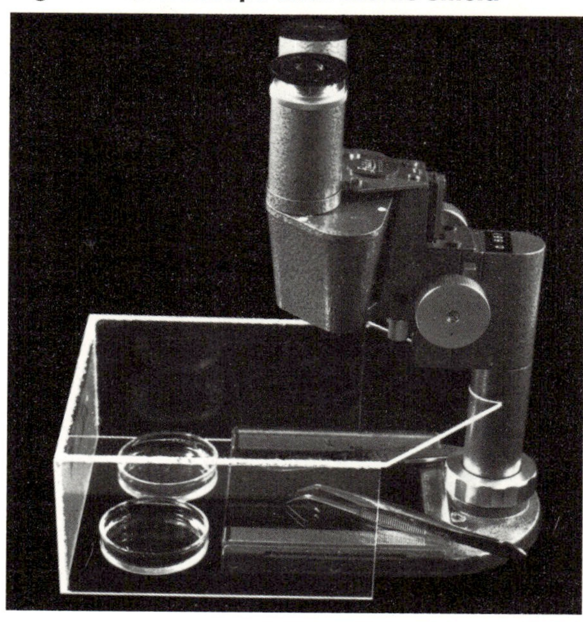

Experimental Procedures

Remove the 'pseudo bulbs' from the plant (Fig. 10.3A and B) taking care to separate all the old leaves and any necrotic tissue until the 'side buds' (Fig. 10.4C) which contain the 'meristems' are revealed (arrowed Fig. 10.3C). At this stage the 'meristems' are still covered with small leaves. Cut through the base of each 'side bud' with a scalpel and dip each piece into a beaker of 95% ethanol for 10 s, remove, shake off excess ethanol and immerse in the hypochlorite solution, also

Fig. 10.3 Location of side buds

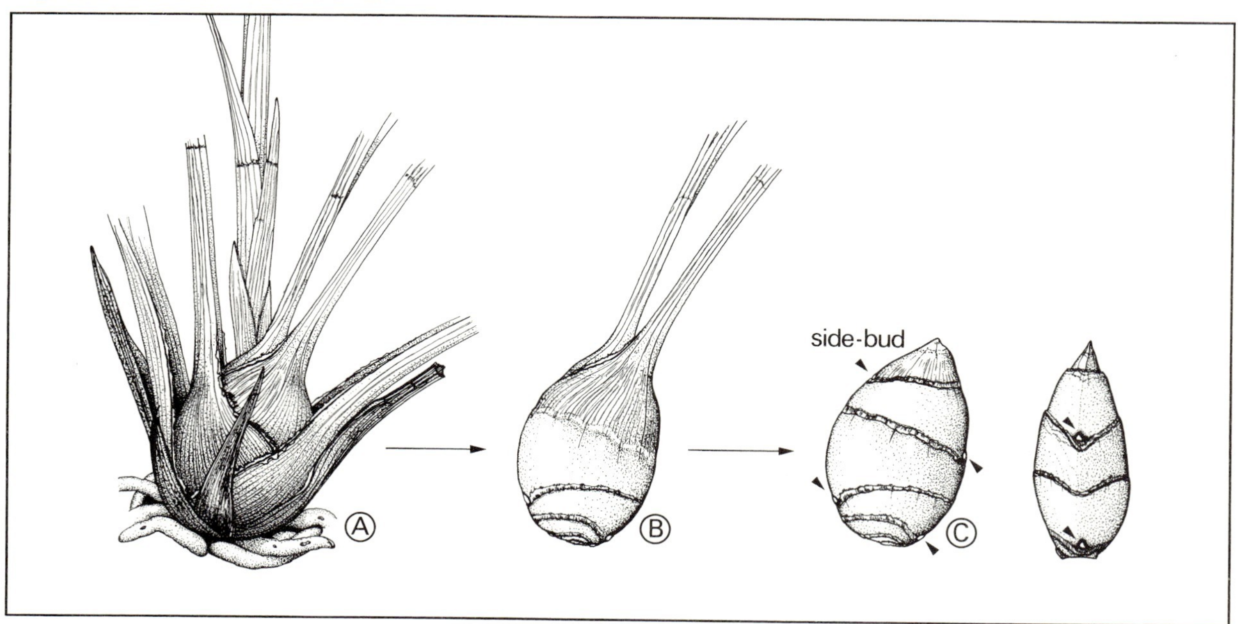

Fig. 10.4 Position of explant (L.S. side bud)

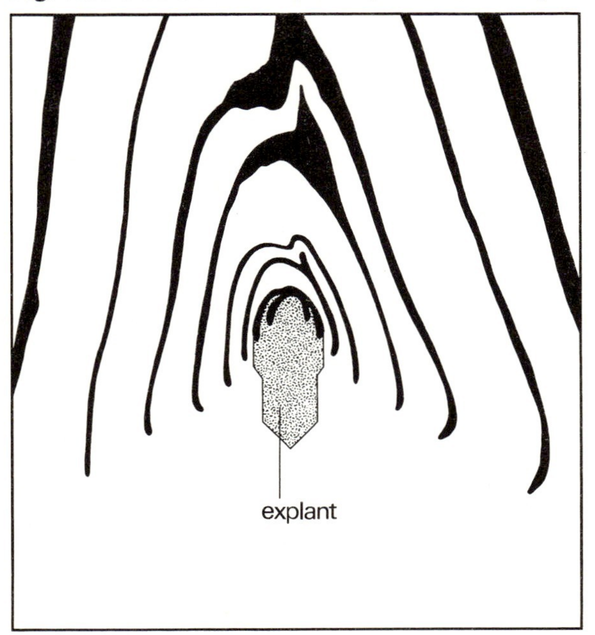

contained in a 250 ml beaker, for 25 min. Transfer the material, still in the sterilising solution, to the sterile room. After sterilisation remove from the hypochlorite and pass successively through three changes of sterile distilled water contained in 250 ml Erlenmeyer flasks. This treatment will remove all traces of the sterilant. Place each bud in a sterile 50 mm Petri dish and carefully remove the young leaves and scale leaves, dropping the bud briefly (10 s) into 95% ethanol, after removing each leaf, and then rinsing off the ethanol in sterile water. Leave the last 2–3 leaf primordia associated with the meristem and transfer the piece of tissue to a 50 mm Petri dish. Next remove the remaining primordia with a very sharp scalpel under a dissecting microscope. Ensure that the glass stage of the microscope and the Plexiglas shield are thoroughly sterilised with 70% ethanol before the removal of the meristem. Carefully separate the tiny 'meristem' (Fig. 10.4) from the rest of the tissue and transfer it to the surface of the agar medium in a culture tube. Incubate the cultures at 25 °C in continuous white light with an irradiance of 2,000 lux. If the explanted meristem is undamaged it will turn green after approximately 8 days and develop 2 weeks later into a perfect protocorm. After these protocorms have developed, sub-divide them within the culture tube using the following aseptic procedure. Push the protocorm off the agar on to the side of the glass tube and cut into 4–6 pieces with a sharp sterile scalpel. Then return the pieces to the surface of the agar medium, flame the tube, and close with aluminium foil. Each piece will now develop rapidly into a new protocorm which can be subdivided again to produce a fresh crop of protocorms.

Scheduling

Event	Timing
Removal of 'meristems' and initiation of culture at 25 °C in continuous light with an irradiance of 2,000 lux	Day 0
Meristem, if undamaged, turns green	c. Day 8
Protocorm fully developed	c. Day 29
Sub-division of protocorm into 4–6 pieces	c. Day 29
Appearance of a fresh crop of fully developed protocorms which can be further sub-divided and sub-cultured	c. Day 50

Recording of Results

Copy down the details of the experiment into a notebook recording the starting date of the experiment, the duration, number of contaminated cultures, and the treatment employed. Make visual observations and drawings of the cultures at regular intervals recording the changes in the cultures.

Questions and Comments

How many orchid plants can under ideal conditions be obtained from a single isolated meristem in one year?

Can you suggest why only some orchid species respond to this culture procedure?

References

Holdgate, D. P.: Propagation of ornamentals by tissue culture, pp. 18–42. In: Applied and Fundamental Aspects of Plant Cell, Tissue and Organ Culture (Reinert, J., Bajaj, Y. P. S. eds.). Berlin-Heidelberg-New York: Springer 1977

Morel, G.: Eine neue Methode erbgleicher Vermehrung. Die Kultur von Triebspitzen-Meristemen. Orchidee *16*, 165–176 (1965)

Rao, A. N.: Tissue culture in the Orchid industry, pp. 44–65. In: Applied and Fundamental Aspects of Plant Cell, Tissue and Organ Culture (Reinert, J., Bajaj, Y. P. S. eds.). Berlin-Heidelberg-New York: Springer 1977

V Isolation, Culture and Fusion of Protoplasts from Higher Plants

V Isolation, Culture and Fusion of Protoplasts from Higher Plants

Experiment 11 Isolation and Culture of Mesophyll Protoplasts from Tobacco Leaves

Isolated plant protoplasts are cells from which the wall has been removed either mechanically or enzymically. The potential of cells without walls as tools to study cellular phenomena was appreciated by cell biologists as long ago as the end of the last century, but it was not until Cocking in 1960 showed that the wall of the constituent cells of the tomato fruit could be removed effectively by cellulase that sufficient material could be generated for experimental purposes. Much earlier, in 1936, Levitt had demonstrated that if cells from onion scale leaf epidermis are cut with a razor blade during strong plasmolysis in 20% sucrose and the cells were then transferred to a weak solution of sucrose, the protoplasts (protoplasm and vacuole) may actually swell enough to come completely out of the wall. Nowadays there are numerous reports of the successful isolation, culture and manipulation of protoplasts from a wide range of higher and lower plants. Leaves have proved to be the most amenable subjects for these procedures, particularly leaves of *Nicotiana tabacum* and *Petunia hybrida*. Cell cultures especially of *Daucus carota* also provide a reasonably easy entry into the arts of protoplast isolation and culture. Isolated protoplasts from all of these species can be induced to divide in culture and eventually to form callus tissue from which plants can be regenerated. In the following experiment, protoplasts will be prepared from the mesophyll tissue of fully expanded leaves of *Nicotiana tabacum* and brought into culture. This provides a beginning to a series of exacting techniques which are of interest to a wide range of biologists.

Materials and Equipment
Sterile Items
(For sterilisation procedures and composition of media see Appendix)

1 One filtration apparatus (Millipore or Sartorius) with filters (0.45 µm) for the sterilisation of enzyme solutions
2 3 Erlenmeyer flasks, 500 ml, each containing 400 ml of distilled water
3 1 Erlenmeyer flask, 250 ml, containing 200 ml of 10% mannitol with inorganic salts CPW (see Appendix)

Fig. 11.1 Items for the sterile transfer room

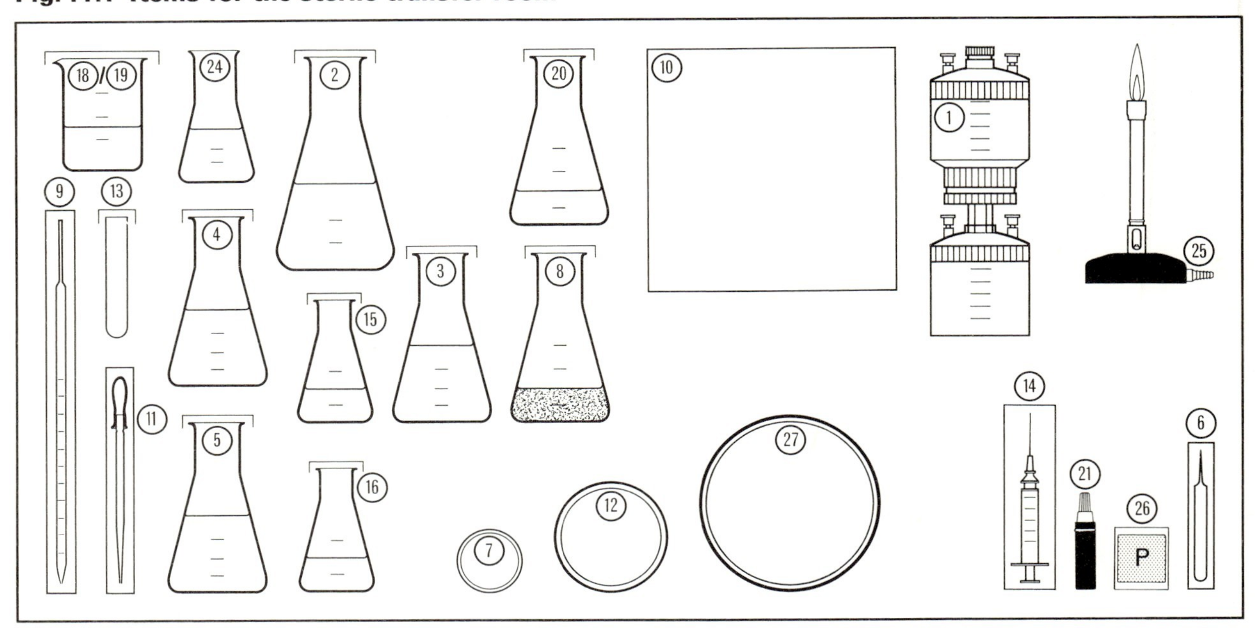

4 1 Erlenmeyer flask, 250 ml, containing 200 ml of 13% mannitol with inorganic salts CPW (see Appendix)
5 1 Erlenmeyer flask, 250 ml, containing 200 ml 20% sucrose with inorganic salts (CPW)
6 1 pair of watchmaker's forceps
7 12 Petri dishes 35 mm in diameter, Falcon Plastics or equivalent
8 1 Erlenmeyer flask, 250 ml, containing 100 ml of M_{PROT} medium containing 1.2% agar at 45 °C
9 3 graduated pipettes 5 ml
10 24 sheets of tissue paper (200×200 mm)
11 6 Pasteur pipettes with rubber or plastic teats
12 6 Petri dishes 90 mm in diameter, glass or plastic
13 6 centrifuge tubes (16 ml) with screw caps
14 1 syringe (10 ml)
15 1 Erlenmeyer flask, 100 ml, containing 50 ml of a mixture of 0.5% cellulase Onozuka-R10 and 0.1% Macerozyme-R10 in 13% mannitol at pH 5.8 (Enzyme solution)
16 1 Erlenmeyer flask, 100 ml, containing 50 ml M_{PROT} medium

Non-Sterile Items

17 6 plants of *Nicotiana tabacum* ('White Burley') 55–65 days after germination
18 250 ml of a solution of sodium hypochlorite approximately 20% v/v. Commercial bleach preparations such as Ace or Domestos can be used at a dilution of 1:5 with water. Add two drops of Tween or similar detergent
19 1 glass beaker (500 ml)
20 1 Erlenmeyer flask (250 ml) containing 100 ml of 70% ethanol
21 1 waterproof marking pen
22 1 low speed centrifuge (50–100×g)
23 1 water bath at 45 °C
24 1 Erlenmeyer flask (150 ml) containing 100 ml 95% ethanol
25 1 Bunsen or ethanol burner
26 1 roll of parafilm
27 1 Petri dish 140 mm in diameter

The items for the sterile transfer room are laid out as shown in Fig. 11.1.

Experimental Procedures

Remove the young fully expanded leaves from the upper part of the plant and immerse in 150 ml of 70% ethanol in a Petri dish (140 mm). After 1 min pour off the ethanol and replace with 150 ml sterile water. After 2 min replace the water with 150 ml of the hypochlorite solution containing 2 drops of Tween 20. After 15 min remove the leaves from the sterilant, wash 3 times in sterile water and blot dry with sterile tissue paper. Remove the lower epidermis from the sterilised leaves as completely as possible with a pair of watchmaker's forceps (Fig. 11.2) and place the leaves, lower surface down, onto 20 ml of a solution of 13% mannitol and inorganic salts (CPW) contained in Petri dishes (90 mm) for 1 h *(preplasmolysis)*. Remove the mannitol-inorganic salts solution with a Pasteur pipette and replace it with the sterile enzyme mixture in mannitol (20 ml). Incubate (overnight) in the dark at 20–22 °C for 16 h and then gently agitate the leaves with sterile forceps to facilitate the release of the protoplasts, pushing the larger pieces of leaf material to one side and keeping the Petri dish at an angle of c. 15° (Fig. 11.3). After 60 min the protoplasts will settle to the bottom of the dish and the enzyme-protoplast

Fig. 11.2 Removal of epidermis

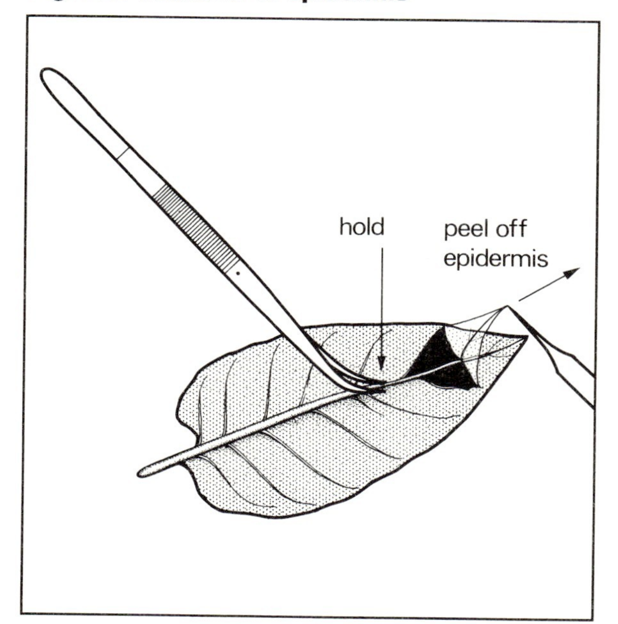

Fig. 11.3 Release of protoplasts

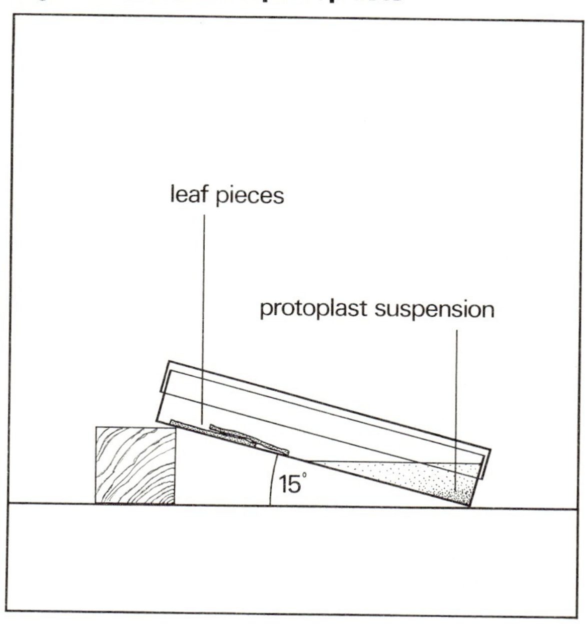

mixture can be transferred using a Pasteur pipette to 15 ml screw capped centrifuge tubes. Centrifuge at 50×g for 10 min and resuspend the pellet in CPW+20% sucrose (Fig. 11.4A) and centrifuge again for 10 min at 50×g (Fig. 11.4B). The viable protoplasts will float to the surface of the sucrose solution while the remaining cells and debris will sink to the bottom of the tube (Fig. 11.4C). Remove the protoplasts with a Pasteur pipette and resuspend in the inorganic salts (CPW)+10% mannitol (Fig. 11.4E), centrifuge again at 50×g for 10 min in order to separate the contaminating debris (Fig. 11.4F and G). Repeat this procedure three times (3×).

Finally, transfer the protoplasts (1.5 ml) with a Pasteur pipette into 1.5 ml of M_{PROT} medium in Falcon Petri dishes (35 mm) at a density of 5×10^4/ml. Seal the dishes with parafilm, culture in the dark for 24 h at 28 °C, then for 2 days with an irradiance of 500 lux and finally for the rest of the experiment at 2,000 lux. The best photoperiod is a 16 h day. The protoplasts should form cell walls and begin to divide after 5 days. Protoplasts may also be cultured in an agar medium by mixing 1.5 ml of the protoplast suspension (10^5/ml) with an equal volume of 1.2% agar medium (M_{PROT}) at 45 °C in a 35 mm Falcon Petri dish. Light and temperature regimes as for the 'liquid' cultures.

The cultures will continue to grow in the media and the first colonies will become visible after 3–4 weeks. At this stage transfer the colonies to new medium with a reduced mannitol level (0.2 M). Protoplasts of cells isolated from carrot cultures behave similarly and can be prepared according to the general procedure for mesophyll protoplasts, however, certain slight modifications must be made to the protocol, i.e., the enzyme concentration must be doubled and a concentration of 15% sucrose must be used to float the protoplasts.

Scheduling

Event	Timing
Sterilisation of leaves; preplasmolysis; treatment of leaves with enzyme solution for 16 h	Day 0
Washing of protoplasts; transfer to medium and culture in the dark for 24 h	Day 1
First transfer of cultures to low light (500 lux) for a 24 h period	Day 2
Second transfer to stronger light (2,000 lux) for a 48 h period	Day 3
Start of prolonged culture	Day 5

Recording of Results

Copy down the details of the experiment in a notebook, recording the starting date and duration of the experiment, and number of cultures. Make visual observations of the cultures recording changes in the morphology and the beginning of cell wall formation and cell division.

Questions and Comments

How may the viability of protoplasts in a culture be determined with certainty?

Which events, apart from mitosis, are essential for the division of protoplasts?

Protoplasts of cells from tobacco leaves as well as those from *Petunia* are able to proliferate and form a callus and to regenerate plantlets after prolonged cultivation under appropriate conditions.

Fig. 11.4 Isolation of protoplasts

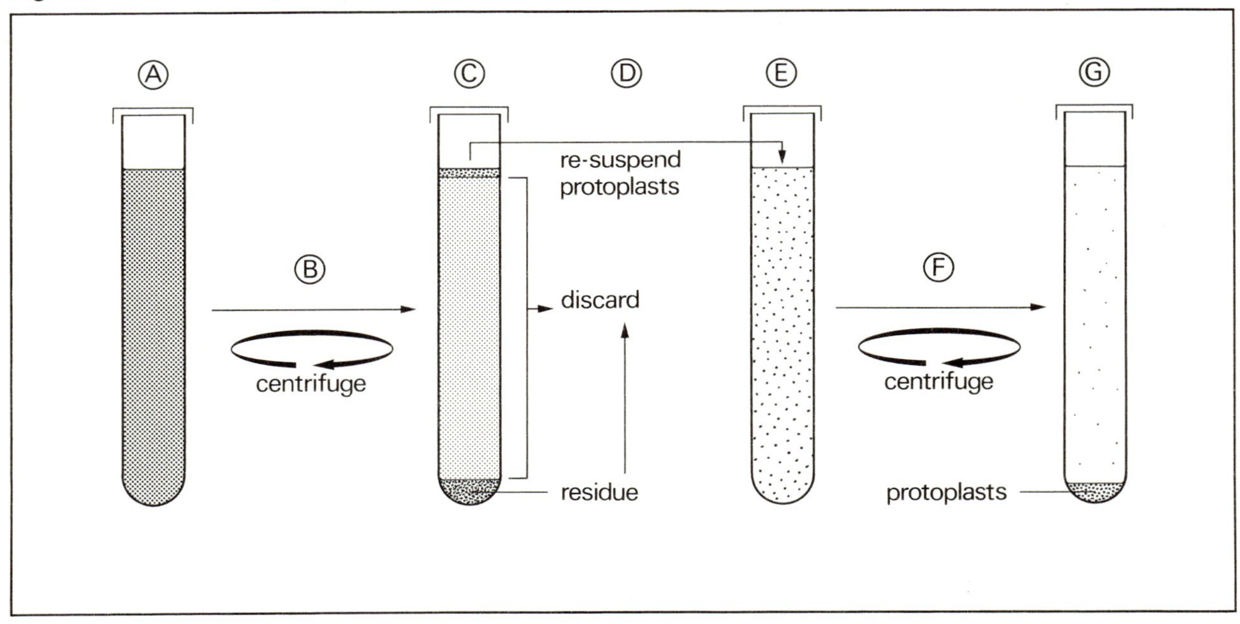

References

Bajaj, Y. P. S.: Protoplast isolation, culture and somatic hybridisation, pp. 467–495. In: Applied and Fundamental Aspects of Plant Cell, Tissue and Organ Culture (Reinert, J., Bajaj, Y. P. S. eds.). Berlin-Heidelberg-New York: Springer 1977

Cocking, E. C.: A method for the isolation of plant protoplasts and vacuoles. Nature (London) *187*, 927–929 (1960)

Cocking, E. C.: Plant cell protoplasts, isolation and development. Annu. Rev. Plant Physiol. *23*, 29–50 (1972)

Levitt, J., Scarth, G. W., Gibbs, R. D.: Water permeability of isolated protoplasts in relation to volume change. Protoplasma *26*, 237–248 (1936)

Nagata, T., Takebe, I.: Plating of isolated tobacco mesophyll protoplasts on agar medium. Planta *99*, 12–20 (1971)

Experiment 12 Protoplast Fusion Induced by Polyethylene Glycol (PEG)

The fusion of protoplasts from different plants to form somatic hybrid cells and the subsequent regeneration of plants from the callus tissue produced from the proliferation of the fusion product was an extremely important achievement. The ultimate aim of this work is to produce hybrids which cannot be produced by normal sexual means. However, isolated protoplasts will not aggregate and fuse easily in the absence of an inducing agent. Several chemicals have been used successfully to induce fusion but the most successful and the one most widely used at the present time is polyethylene glycol (PEG). The effects of PEG are not specific and it will promote the aggregation and fusion of protoplasts from the same or different species. The process of aggregation and fusion can be best observed when the two protoplast populations are quite different in appearance, for example when one contains anthocyanin in the vacuole (some strains of *Daucus carota* cell cultures) and the other contains a number of green chloroplasts (*Nicotiana tabacum* mesophyll cells). Here the hybrid cells will contain both the visible markers and the process of mixing can be observed with ease. In the following experiment mesophyll protoplasts from the leaves of *Nicotiana tabacum* will be aggregated and fused together using PEG.

Materials and Equipment
Sterile Items
(For sterilisation procedures and composition of media see Appendix)

1. 4–6 ml of a freshly isolated protoplast suspension from mesophyll protoplasts of tobacco leaves in M_{PROT} medium at a density of 5×10^5/ml which have been floated and washed thoroughly (see Expt. 11)
2. 50 Pasteur pipettes
3. 10 cover glasses 24×32 mm
4. 12 Petri dishes, 35 mm in diameter, Falcon plastics or equivalent
5. 1 Erlenmeyer flask (50 ml) containing 10 ml of a solution of a Polyethylene glycol 1,500 solution (P_S) (containing 4.5 g PEG, 15.5 mg $CaCl_2$, 0.9 mg KH_2PO_4 at pH 6.2)

Fig. 12.1 Items for the sterile transfer room

6 1 Erlenmeyer flask (100 ml) containing 50 ml CPW+10% mannitol
7 1 Erlenmeyer flask (100 ml) containing 50 ml M_{PROT} medium
8 Silicone oil (5 ml) sterilised by filtration

Non-Sterile Items

9 1 inverted microscope
10 1 compound microscope
11 1 Fuchs-Rosenthal counting chamber with cover-slips
12 1 roll of parafilm
13 1 waterproof marking pen

The items for the sterile transfer room are laid out as shown in Fig. 12.1.

Experimental Procedures

Using a Pasteur pipette place a small drop (ca. 50 µl) of sterile silicone oil on to the centre of a plastic Petri dish (35 mm) and gently lower a cover glass onto the drop of silicone fluid. This will ensure that the cover glass is fixed in the Petri dish. Pipette ca. 150 µl of protoplast suspension directly onto the middle of the cover glass and allow the protoplasts to sink undisturbed to the bottom and stick to the glass (c. 5 min). Now using a Pasteur pipette add 450 µl of the PEG solution drop by drop to the outer edge of the protoplast culture, placing the last drop in the centre. This will ensure that the protoplast suspension and the PEG will mix without undue disturbance. After 20–40 min at room temperature (c. 22 °C) carefully add drop by drop to the periphery of the mixture 500 µl of CPW solution+10% mannitol. Repeat this treatment again after 10 min and again after a further 10 min, but this time with *1.5 ml* of CPW+10% mannitol. Follow this protocol exactly! Five minutes after the last wash carefully suck off the solution with a Pasteur pipette leaving the protoplasts with a thin film of medium. In these circumstances the fusion products which will stick to the cover glass may then be washed three times with 1 ml aliquots of CPW+10% mannitol as before. Aggregation begins during this washing procedure and can be best observed c. 60 min after the start of culture which is initiated by the addition of 1 ml of M_{PROT} medium. At this stage the protoplasts and fusion products are set free from the coverslip and can be suspended in the medium using a Pasteur pipette. Replace the lid of the Petri dish and seal with parafilm.

Incubate the cultures initially in the dark for 24 h and then transfer to white light (1,000 lux) at 28 °C. The fusion products can be easily observed using an inverted microscope.

Scheduling

Event	Timing
Preparation of protoplasts on cover slips and treatment with PEG solution	Day 0
Washing of protoplasts	Day 0
Transfer to culture medium	Day 0
Culture in the dark at 28 °C	Day 1
Culture in light (1,000 lux) at 28 °C	Day 2

Recording of Results

Copy down the details of the experiment into a notebook, recording the starting date and duration of the experiment, the media and culture conditions. Make drawings of the isolated protoplasts and of fusion products.

Questions and Comments

What would be the best 'marker' for the identification of fusion products?

What would be the reason that protoplasts will not aggregate and fuse without PEG?

References

Eriksson, T.: Technical advances in protoplast isolation and cultivation, pp. 313–322. In: Plant Tissue Culture and its Bio-technological Application (Barz, W., Reinhard, E., Zenk, M.H. eds.). Berlin-Heidelberg-New York: Springer 1977

Gamborg, O.L.: Somatic cell hybridization by protoplast fusions, and morphogenesis, pp. 287–301. In: Plant Tissue Culture and its Bio-technological Application (Barz, W., Reinhard, E., Zenk, M.H. eds.). Berlin-Heidelberg-New York: Springer 1977

Gosch, G., Bajaj, Y.P.S., Reinert, J.: Isolation, culture and fusion studies on protoplasts from different species. Protoplasma *85*, 327–336 (1975)

Kao, K.N., Michayluk, M.R.: A method for high-frequency intergeneric fusion of plant protoplasts. Planta *115*, 355–367 (1974)

VI Secondary Metabolites in Tissue Cultures

VI Secondary Metabolites in Tissue Cultures

Experiment 13 Callus Formation and Anthocyanin Production in Cultures of *Haplopappus gracilis*

Cultures of higher plant cells can accumulate a variety of secondary metabolites such as polyphenols, alkaloids, steroids or certain pigments. Some of these secondary products can accumulate in the medium and may inhibit growth. This is particularly true of polyphenols. Other compounds, anthocyanins for instance, are not secreted by viable cells. Anthocyanins are a group of red, blue or violet pigments which occur in the vacuolar sap of plant cells, often in the epidermis of flowers and fruits. The products of hydrolysis of an anthocyanin are an aglycone bearing the generic name anthocyanidin and a glycoside unit, often glucose. Cultures which possess the ability to synthesise and accumulate pigments tend to lose this property during prolonged cultivation. This appears to be a phenomenon similar to the loss of morphogenetic capability with ageing of tissue cultures. Cells which can synthesise and accumulate pigments provide useful material in a practical course because the appearance and accumulation of the pigment can be monitored visually, as for example, with anthocyanin. In *Haplopappus* cultures anthocyanin formation is induced by light. It is also possible to demonstrate that different parts of *Haplopappus* seedlings exhibit different abilities for callus formation and apparently different potentials for anthocyanin accumulation.

Materials and Equipment
Sterile Items
(For sterilisation procedures and composition of medium see Appendix)

1. 6 established friable *Haplopappus* cultures, at least 3 months since initiation, growing in the dark on M_{HAP} agar medium with 5×10^{-9} g/ml 2,4-D, at 23 °C
2. 10 culture tubes (150×25 mm) containing 10 ml of M_{HAP} agar medium without 2,4-D, sloped at an angle of 25°
3. 20 culture tubes (150×25 mm) containing 10 ml M_{HAP} agar medium with 2,4-D. sloped at an angle of 25°
4. 3 Erlenmeyer flasks (100 ml) each containing 80 ml of water
5. 5 Petri dishes, 90 mm in diameter, glass or plastic
6. 1 Erlenmeyer flask (100 ml) containing 50 ml liquid M_{HAP} medium

Formulae of anthocyanins and anthocyanidins:

Anthocyanins:

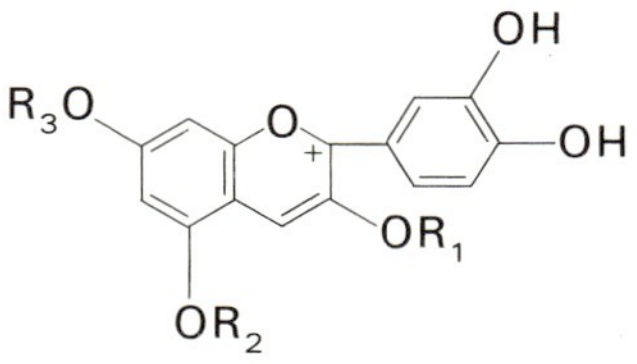

R$_3$O, O, +, OH, OH, OR$_1$, OR$_2$

Free base (pH 7–8): $R_1 = R_2 = R_3 = H$

Monoglucoside: R_1 = Glucose, $R_2 = R_3 = H$

Disaccharide: $R_3 = H$ $R_1 = R_2$ = Glucose

$R_2 = H$ $R_1 = R_3$ = Glucose

Anthocyanidins:

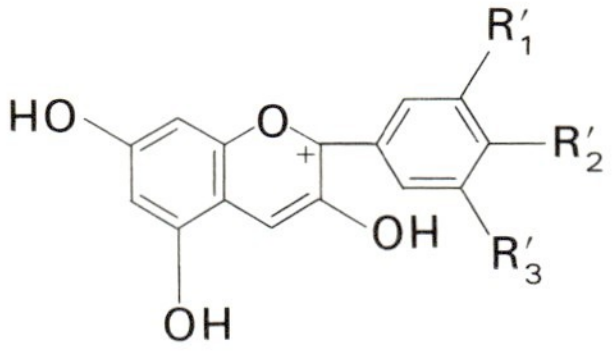

HO, O, +, OH, OH, R'_1, R'_2, R'_3

Pelargonidin: $R'_1 = R'_3 = H$, $R'_2 = OH$

Cyanidin: $R'_1 = R'_2 = OH$, $R'_3 = H$

Malvidin: $R'_1 = R'_3 = OCH_3$, $R'_2 = OH$

Non-Sterile Items

7 150 seeds of *Haplopappus gracilis*
8 2 racks, metal or plastic, each to hold 12 culture tubes (150×25 mm)
9 6 Pasteur pipettes, 40 mm in length, each with a rubber teat
10 Cotton wool
11 1 glass beaker (100 ml) containing 90 ml of Ace or Domestos solution at a dilution of 1:5 with water to which one drop of detergent has been added
12 1 compound microscope
13 1 platinum wire mounted in a metal handle
14 1 Erlenmeyer flask (150 ml) containing 100 ml 95% ethanol
15 1 Bunsen or ethanol burner
16 1 waterproof marking pen

The items for the sterile transfer room are laid out as shown in Fig. 13.1.

Experimental Procedures

Select approximately 100 of the larger seeds. *Haplopappus* seeds are small with stiff hairs and are difficult to handle when wet. In order to simplify the handling of the seeds the following special technique should be used.

Take a Pasteur pipette without the rubber teat and block up the narrow end with a small piece of cotton wool. Place 30–40 seeds on top of the cotton wool layer inside the pipette and then close up the wide opening of the Pasteur pipette with another piece of cotton wool, finally replacing the rubber teat (Fig. 13.2). Now place the tip of the Pasteur pipette into the hypochlorite solution and suck up the sterilising agent into the body of the pipette through the cotton wool. Take care that all of the air bubbles have disappeared when the Pasteur pipette is filled up with hypochlorite as the presence of air bubbles will seriously affect the success of the sterilisation procedure. After the seeds have been in contact with the hypochlorite 30 min, expel the hypochlorite solution using the teat and replace it with sterile distilled water. Repeat this procedure half a dozen times to ensure that all of the hypochlorite has been removed from the surface of the seeds. Repeat the complete procedure three times.

Fig. 13.2 Pipette for handling seeds

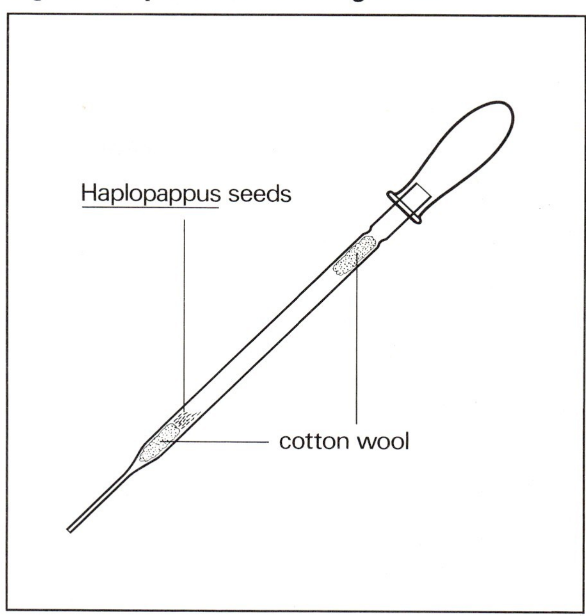

Fig. 13.1 Items for the sterile transfer room

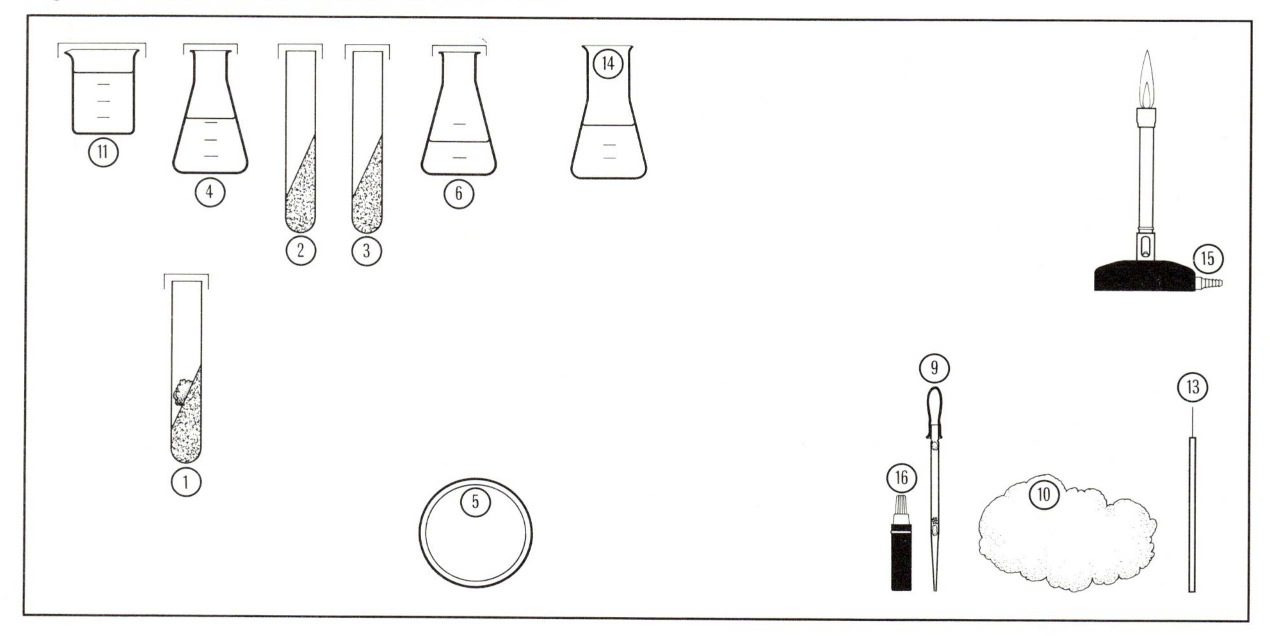

Remove the rubber teat from a Pasteur pipette containing the sterile seeds and flame the wide opening carefully and remove the cotton wool plug. Insert the flamed platinum wire into the tip of the pipette, some of the tiny seeds will stick to the wire which can then be withdrawn and transferred to the surface of the culture medium (M_{HAP} without 2,4-D). Ensure that the seeds are evenly distributed over the agar surface. Flame, seal with aluminium foil, mark, place in a rack and incubate in white light (5,000 lux) for 2 days and subsequently in darkness at 26 °C for 7 days. Examine each culture after 4 days and discard any infected culture tube.
After 9 days the seedlings will have grown to a height of 15–30 mm. Only use individuals of this size and reject all seedlings which are smaller or larger.
Take three tubes containing M_{HAP} medium with 2,4-D and one tube (with seedlings) which has been flamed, opened, flamed again and allowed to cool. Gently pull one seedling to the upper edge of the medium and cut it into three pieces (stem-meristem+cotyledons, stem+hypocotyl, and root), starting either from the cotyledonary node or the root tip. Transfer these explants to a tube of fresh medium (M_{HAP} with 2,4-D) and place these spatially in sequence, as in the seedling, on the agar surface. Incubate in the dark at 26 °C. After 3 weeks callus will have been produced in most of the explants and the potential of different parts of the seedling for callus formation can be determined. Callus can now be removed and used to initiate stock cultures on M_{HAP} medium in the dark at 23 °C for the experiments on anthocyanin synthesis and accumulation.
Anthocyanin formation is rather weak in young cultures (4–8 weeks old). Therefore cultures at least 3 months old and pigment-free after growth in the dark must be used. Transfer two pieces of callus from the actively growing parts of the established cultures to each of the 10 tubes containing M_{HAP} with 2,4-D. Spread the soft friable callus cautiously on the surface of the agar medium. This is necessary to secure a sufficiently large area for pigment production which is restricted to the upper cell layers. After sealing and marking the culture tubes, place in a rack and expose to white light (Osram 25, Universal fluorescent tube, 5,000 lux or Wotan HQI-TS 400 W/D with a blue filter) or to blue light (Phillips 40 W/15, 5,000 ergs/cm^2/s=19.2 µE/m^2/s =c 430 nm, 5,000 ergs/cm^2/s, but not less than 2,500 ergs) for 6 days at 23 °C. Some of the cultures will synthesise and accumulate anthocyanin but the extent of pigment formation will depend upon the original position of the explant in the seedling. The extent of pigment formation can be examined qualitatively after removing and spreading small samples of cells on slides in liquid M_{HAP} medium under the microscope.

Scheduling

Event	Timing
Initiation of sterile seedlings	Day 0
Examine and discard contaminated cultures	Day 4
Initiation of callus cultures	Day 9
Isolation and first transfer of *Haplopappus* callus	Day 30
Initiation of pigment formation in established cultures (3 months old) and beginning of the light treatment	Day 0
Anthocyanin formation	Day 7

Recording of Results

Copy down the details of the experiments into a notebook recording the starting date and duration of the experiment, media and light treatments employed. Examine pigmented cultures microscopically.

Questions and Comments

Do all cells of the *Haplopappus* cultures form anthocyanins?
What conclusions can be drawn about the nature of the receptor from the fact that blue light is more effective in anthocyanin formation than other wave lengths?
Older *Haplopappus* seeds are sometimes difficult to germinate. It is advisable therefore to use only last year's crop.

References

Geissman, T.A.: Anthocyanins, Chalcones, Aurones, Flavones, and Related Water-Soluble Plant Pigments, pp. 450–498. In: Moderne Methoden der Pflanzenanalyse III. (Paech, K., Tracey, M.V. eds.). Berlin-Heidelberg-New York: Springer 1955

Grill, R., Vince, D.: Anthocyanin formation in turnip seedlings (*Brassica rapa* L.). Evidence for two light steps in the biosynthetic pathway. Planta *63*, 1–12 (1965)

Reinert, J., Torrey, J. G.: Über die Kultur von Geweben aus *Haplopappus gracilis*. Naturwissenschaften *48*, 132–133 (1961)

Reinert, J., Clauss, H., Ardenne, von R.: Anthocyanbildung von *Haplopappus gracilis* in Licht verschiedener Qualität. Naturwissenschaften *51*, 87 (1964)

Experiment 14 The Production of the Steroid, Diosgenin, from Tissue Cultures of *Dioscorea deltoidea*

Experiment 14 The Production of the Steroid, Diosgenin, from Tissue Cultures of *Dioscorea deltoidea*

There are few plant tissue and cell cultures which can produce levels of secondary products comparable with that of the intact plant. However, notable exceptions are callus and suspension cultures derived from *Dioscorea deltoidea* which can accumulate the steroid diosgenin up to a level exceeding 1% of the dry weight. Further, the yield can be increased substantially by feeding the cultures with the precursor cholesterol.

The aim of this experiment is to demonstrate the accumulation of the steroid diosgenin and related compounds in callus cultures of *Dioscorea deltoidea*. The techniques of extraction, separation and detection of diosgenin in thin-layer chromatograms are also described.

Materials and Equipment

Sterile Items

(For sterilisation procedures and composition of medium see Appendix)

1 Three callus cultures of *Dioscorea deltoidea* initiated from seedlings and previously subcultured at least 5 times
2 9 wide necked Erlenmeyer flasks (250 ml) containing 100 ml of culture medium solidified with 1% agar (M_{DIO})
3 25 sheets of aluminium foil (100×100 mm)
4 10 Petri dishes, 90 mm in diameter, either glass or plastic
5 3 scalpels (150 mm)
6 1 pair of large forceps (250 mm)

Non-Sterile Items

7 9 foil cups (60 mm in diameter) for drying and weighing tissue
8 4 Pyrex round bottomed flasks (250 ml) for Soxhlet extractor
9 Buchner funnel to take 11 cm filter paper
10 2 Pyrex Soxhlet extractors (siphoning volume 60 ml)
11 4 Pyrex condenser units
12 9 cellulose thimbles (Whatman) (25×80 mm)
13 1 rotary evaporator
14 Sample bottles (10 ml)
15 1, 10 μl Hamilton syringe
16 2 glass T.L.C. plates (200×200 mm)

Fig. 14.1 Items for the sterile transfer room

17 1 Shandon Chromotank
18 1 reagent spray gun (glass)
19 Chemicals: 10% HCl, chloroform, chromatography standards (diosgenin, cholesterol, sitosterol and stigmasterol), T.L.C. solvent (benzene: ethyl acetate 3:1), Anisaldehyde reagent
20 10 filter paper circles (110 mm) Whatman No. 52 (hardened)
21 1 desiccator
22 1 hot air drying oven
23 2 l of distilled water
24 1 analytical balance
25 1 Bunsen or ethanol burner
26 1 Erlenmeyer flask (150 mm) containing 100 ml 95% ethanol

The items for the sterile transfer room are laid out as shown in Fig. 14.1.

Experimental Procedures

Take one 250 ml flask containing a *Dioscorea* callus culture and lightly flame the neck before discarding the foil cap, then flame the neck again before removing the callus with sterile forceps and place the tissue in a sterile Petri dish. Divide the callus into three pieces, removing any necrotic tissue. Ensure that the lid of the Petri dish is replaced after each manipulation. Flame and discard the foil cap from a flask of fresh medium and then flame the neck before placing a piece of callus onto the medium with flamed forceps; seal the flask with a double layer of foil. Repeat this procedure until all the cultures have been transferred. The new cultures should then be placed in a controlled environment room for 6 weeks at 25 °C under continuous low intensity white light.

Place each piece of harvested tissue (10–15 g) in a foil cup and leave in a hot air oven at 90 °C for 24 h. Remove, cool in a desiccator and determine the dry weight. The dried tissue must now be hydrolysed to remove the sugar residues from the diosgenin which is present in the cells as a glycoside or saponin, dioscin (see formula). Place the dried tissue in a 250 ml round bottomed flask, add 10% HCl (10 ml per g fresh weight) and reflux for 4 h, then filter off the tissue using a Buchner funnel and wash with 4×250 ml of distilled water. Dry the tissue overnight at 90 °C before further processing. For the extraction, place the tissue in a cellulose thimble and reflux in a Soxhlet extractor using 80 ml of chloroform. After 8 h remove the chloroform extract from the Soxhlet flask and distil off the chloroform using a rotary evaporator. Redissolve the extract in about 4 ml of chloroform and place this in a sample bottle with a good seal. This is the sample to be assayed by thin-layer chromatography (TLC).

Using a 10 µl Hamilton syringe carefully spot onto a 200×200 mm glass TLC plate (0.2 mm of silica gel) 1 µl, 5 µl and 10 µl of chloroform extract and 10 µl (10 µg) each of the chromatography standards, diosgenin, cholesterol, sitosterol and stigmasterol, made up in chloroform. Place the two spotted plates into a Shandon Chromotank containing 100 ml of solvent (benzene: ethylacetate 3:1); the tank should be allowed to saturate with solvent vapour for 1 h before use. When the solvent has reached a mark 10 mm from the top of the plates, remove and allow to dry. Now spray the plates with Anisaldehyde reagent (50 ml glacial acetic acid, 1 ml concentrated sulphuric acid and 0.5 ml anisaldehyde), and heat in an oven for 5 min at 90 °C. Spots will then appear, which have a characteristic colour, i.e., diosgenin, yellow and sterols, violet.

Formulae of dioscin and diosgenin:

Dioscin

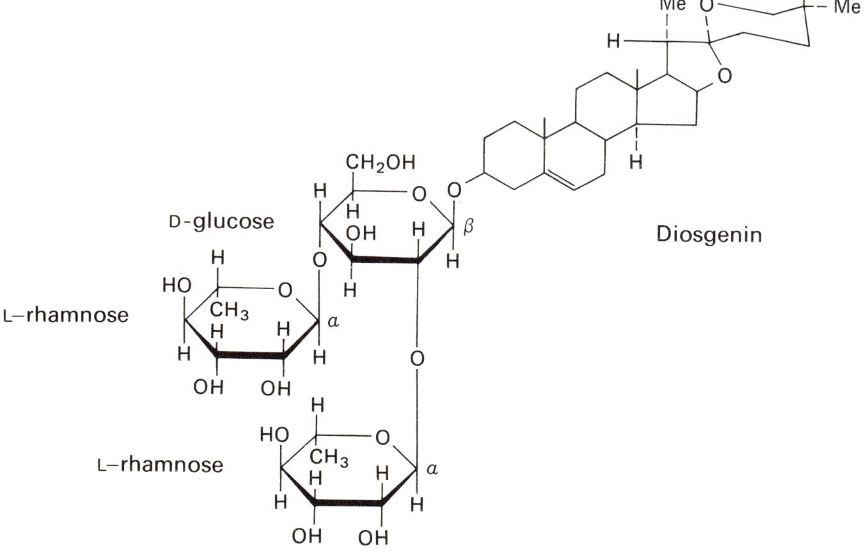

D-glucose

L-rhamnose

L-rhamnose

Diosgenin

Record the colour and pattern of the spots immediately before the colour fades. If the extract is too concentrated and the TLC plates have not run satisfactorily then appropriate changes should be made to the initial volume of the chloroform extract spotted on to the plates.

Scheduling

Event	Timing
Subculture of *Dioscorea deltoidea* callus	Day 0
Harvesting and drying of cultures	c. Day 42
Hydrolysis of dried cultures in 10% HCl	Day 43
Dried, hydrolysed residue extracted with chloroform	Day 44
Concentration and TLC analysis of chloroform extract	Day 45

Recording of Results

Copy down the details of the experiment into a notebook recording the starting date and duration of the experiment, and the number of cultures. Record the number of compounds separated from the callus extract and make a diagram of the chromatogram. Estimate approximately by comparing with the standards of diosgenin present in each culture.

Questions and Comments

How could this qualitative technique using TLC be adopted to provide a quantitative measure of the amounts of the various steroidal components in the extract?

Devise an experiment to demonstrate the accumulation of diosgenin with time and attempt to show the effect of added cholesterol on diosgenin in callus cultures. Can the accumulation of diosgenin be correlated with the growth and differentiation of the callus?

References

Kaul, B., Staba, E.J.: *Dioscorea* tissue cultures. I. Biosynthesis and isolation of diosgenin from *Dioscorea deltoidea* callus and suspension cultures. Lloydia *31,* 171–179 (1968)

Kaul, B., Stohs, S.J., Staba, E.J.: *Dioscorea* tissue cultures. III. Influence of various factors on diosgenin production by *Dioscorea deltoidea* callus and suspension cultures. Lloydia *32,* 347–359 (1969)

VII Embryo and Organ Culture

VII Embryo and Organ Culture

Experiment 15 Embryos of Maize *(Zea mays)*

The removal and culture of embryos of higher plants was one of the earliest successful techniques in plant tissue and organ culture. P. R. White desribed a simple technique in his book on tissue culture in which embryos removed from the seeds of Shepherd's Purse *(Capsella bursa-pastoris)* can be cultured under completely aseptic conditions. Over the years the excised embryos of many species have been brought into culture using relatively simple nutrient media. The advantages of growing an embryo isolated from the rest of the seed, apart from the intrinsic interest in doing so, are to remove the immature plant from the endosperm and/or cotyledon(s) which may in particular cases prevent or modify the development of the plant. This is particularly true, for instance, in the case of sexual hybrids between certain cereals, where the hybrid embryo cannot survive unless removed and cultured in isolation. Studies on the strong metabolic relationships which exist between embryo and endosperm, or embryo and cotyledon(s), are also possible using excised embryos in culture. In certain instances the excised embryo can also be used as a means of propagating species which resist attempts to use standard methods of vegetative propagation. In this experiment the subject, an embryo of maize, is large, easy to remove from the grain, and can be brought into sterile culture easily and successfully. Once the principles have been mastered with a relatively simple example it is so much easier to proceed to other more difficult but perhaps more interesting examples.

Materials and Equipment
Sterile Items
(For sterilisation procedures and composition of medium see Appendix)

1 6 Petri dishes 90 mm in diameter, plastic or glass, containing two layers of filter paper (Whatman No. 1) liberally moistened with distilled water
2 3 Erlenmeyer flasks (250 ml) each containing 200 ml of sterile distilled water
3 6 Petri dishes (plastic) 50 mm in diameter, containing 3 ml of culture medium M_{EMB} (see Appendix)
4 6 Petri dishes 90 mm in diameter, plastic or glass

Fig. 15.1 Items for the sterile transfer room

5 3 scalpels (c. 150 mm)
6 3 pairs of forceps (120–150 mm)

Non-Sterile Items

7 15 grains of maize, *Zea mays*, undamaged and without discoloration
8 1 Pyrex glass tube, 25 mm in diameter and 160 mm in length
9 1 piece of muslin, 60×60 mm
10 2 small rubber bands, c. 15 mm in diameter
11 1 Erlenmeyer flask (150 ml) containing 100 ml 95% ethanol
12 1 Erlenmeyer flask, wide necked (250 ml) containing 100 ml of 95% ethanol
13 1 Erlenmeyer flask, wide necked (250 ml) containing 100 ml of Ace or Domestos at a dilution of 1:5 with water
14 1 Erlenmeyer flask, wide necked (250 ml) containing, 100 ml of 1% Tween 20 (or similar detergent) solution in water
15 1 waterproof marking pen
16 2 plastic or metal trays to hold Petri dishes and Erlenmeyer flasks
17 1 roll of parafilm
18 1 Bunsen or ethanol burner

The items for the sterile transfer room are laid out as shown in Fig. 15.1, preferably with laminar air flow bench.

Experimental Procedures

Take 15 grains and place them in the 'filter tube' (Fig. 15.1). Insert the tube into 100 ml of 95% ethanol contained in a 250 ml Erlenmeyer flask for 10 s, remove, shake off the excess ethanol and place in a second flask containing 100 ml of 1% Tween 20 solution. Shake gently for 1 min and then examine the individual grains noting and removing those with translucent patches beneath the pericarp caused by the penetration of water. Grains with an intact wall will remain unchanged in appearance and can then be surface sterilised by transferring the 'filter' tube to a third flask containing 100 ml of Ace or Domestos for 15 min. During the sterilisation procedure transfer the grains in the flask to the sterile transfer room.

After the sterilisation period is complete take the tube with the grains through two changes of sterile distilled water (150 ml in each 250 ml Erlenmeyer flask) to remove all traces of hypochlorite. Place three grains on top of the moistened filter paper in each 90 mm Petri dish and incubate in the dark for 6 h at 25 °C. During this time the grains will imbibe water and the tissues will become softened. After imbibition, transfer the Petri dishes to the sterile room and remove the embryo from each grain according to the following procedure. Place a sterile grain in a sterile Petri dish and remove the pericarp and testa (fused together) by gently peeling with forceps holding the grain stationary with a second pair of forceps (Fig. 15.2A and B). Remove the endosperm by carefully cutting with a scalpel from the side opposite to the embryo thus revealing the underlying scutellum containing the embryo (Fig. 15.2C). After excision, rinse each embryo and scutellum with sterile distilled water (5 ml) and transfer immediately to a 50 mm Petri dish containing 3 ml of culture medium. Seal each dish with a strip of parafilm and incubate in the dark at 25 °C.

Fig. 15.2 Excision of scutellum and embryo

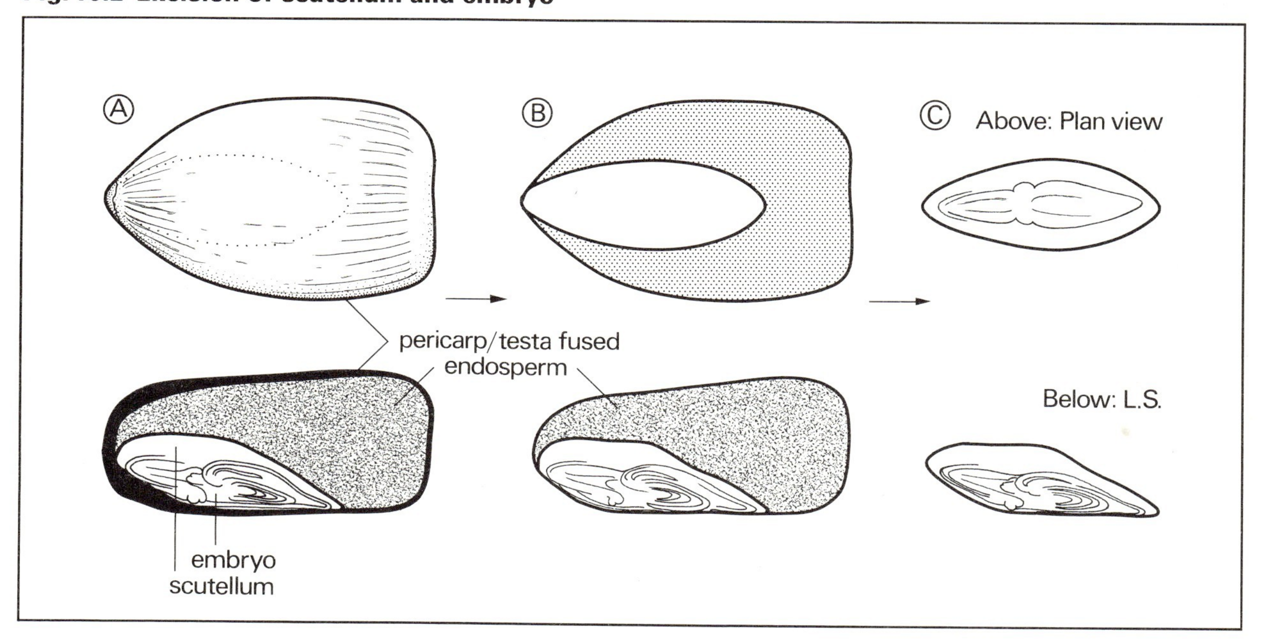

Scheduling

Event	Timing
Imbibition of sterilised grains	Day 0
Surgical removal of the embryos from the intact grain and initiation of culture	Day 0+6 h
Emergence of radicle	Day 2
Emergence of plumule	Day 3–4
Extensive proliferation of the primary root and the emergence of some adventitious roots	Day 7–9

Recording of Results

Copy down the details of the experiment into a notebook recording the starting date of the experiment, the duration, and number of cultures. Make visual observations and drawings of the embryos at approximately daily intervals. Measure the length of the primary root and the number of adventitious roots at each interval. Incorporate your results into a table.

Questions and Comments

Is the rate of development of the excised embryo likely to be different from that of the intact grain?
Give reasons for the variations in growth rate between individual cultures.

References

Cooper, K. V., Dale, J. E., Dyer, A. F., Lyne, R. L., Walker, J. T.: Early development of hybrids between barley and rye. Eucarpia, Proc. VIII Congr. Interspecific Hybridization in Plant Breeding, Madrid, Spain, pp. 275–283 (1977)

Raghavan, V.: Nutrition, growth and morphogenesis of plant embryos. Biol. Rev. *41*, 1–58 (1966)

Rijven, A. H. G. C.: In vitro studies of the embryos of *Capsella bursa-pastoris.* Acta Bot. Neerl. *1,* 157–200 (1952)

White, P. R.: The Cultivation of Plant and Animal Cells, 2nd Ed. New York: Ronald Press 1963

Experiment 16 The Isolation and Culture of the Primary Seedling Root of Pea *(Pisum sativum)*

Experiment 16 The Isolation and Culture of the Primary Seedling Root of Pea *(Pisum sativum)*

The earliest successful sterile cultures of parts of higher plants were started from excised primary seedling roots. Root tips are attractive candidates because of their meristematic nature, availability, and the property of remaining relatively unperturbed after excision. In these early experiments the excised root tips were grown on a solid substratum from which the excised organs derived both nutrients and mechanical support. The media used were fairly simple in composition and contained, in addition to a natural extract, a carbon source, glucose or sucrose, and a mineral salts mixture consisting of the essential macro and micro elements. Although roots were subcultured with some success it was left to P. R. White to show that detached roots could be grown in aseptic culture for unlimited periods. Subsequently, in a series of investigations roots of pea and tomato have been cultivated in vitro and also used to study basic growth phenomena.

Modern methods for the culture of excised root tips are all basically similar. The method used in this experiment depends on the germination of a surface sterilised seed under conditions of strict asepsis. The radicle which emerges is excised and transferred to a liquid medium and incubated without agitation. The theoretical advantage of shaking root cultures is that gaseous and nutrient gradients are eliminated. However, using the technique described below the roots grow at a similar rate whether agitated or not which suggests that aeration and nutrient supplies are adequate. At intervals, and preferably when the root is growing actively, the terminal 10 mm of the root is removed and placed in fresh medium. After two or three subcultures the growth rate slows down unless the medium is supplemented with thiamin and nicotinic acid.

Materials and Equipment

Sterile Items

(For sterilisation procedure and composition of medium see Appendix)

1. 3 Erlenmeyer flasks (250 ml) each containing 200 ml of distilled water
2. 24 Petri dishes 90 ml in diameter, glass or plastic
3. 12 rimless glass culture tubes (150×25 mm) each containing 20 ml distilled water

Fig. 16.1 Items for the sterile transfer room

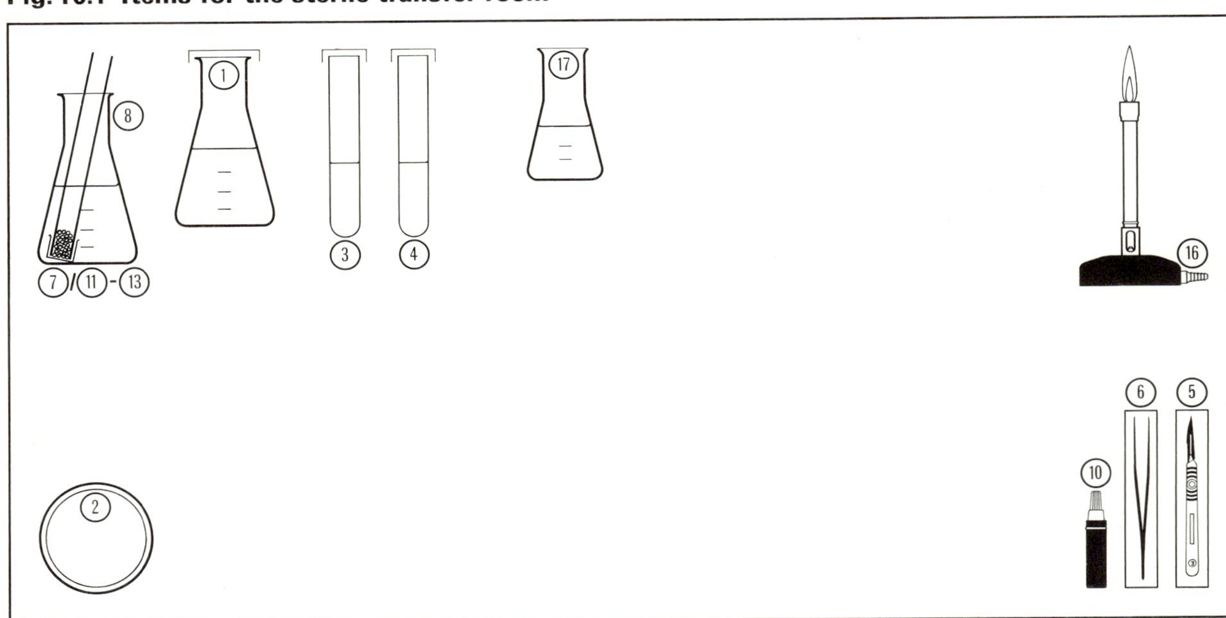

4 12 rimless glass culture tubes (150×25 mm) each containing 20 ml of culture medium M_{ROOT}
5 3 scalpels (150 mm)
6 2 pairs of forceps (120–150 mm)

Non-Sterile Items

7 24 dry, undamaged pea seeds *(Pisum sativum)*, most cultivated varieties will respond in culture
8 200 ml of an aqueous solution of calcium hypochlorite approximately 10% w/v in an Erlenmeyer flask (250 ml)
9 2 Pyrex Erlenmeyer flasks (250 ml) containing 150 ml of 95% ethanol
10 1 waterproof marking pen
11 2 Pyrex glass tubes 25 mm in diameter and 150 mm in length
12 2 pieces of muslin (60×60 mm)
13 2 small rubber bands c. 15 mm in diameter
14 2 plastic or metal trays to hold Petri dishes
15 200 ml of an aqueous solution of 1% Tween 20 (or similar detergent) in an Erlenmeyer flask (250 ml)
16 1 Bunsen or ethanol burner
17 1 Erlenmeyer flask (150 ml) containing 100 ml 95% ethanol

Items for the sterile transfer room are laid out as shown in Fig. 16.1.

Experimental Procedures

Take 24 pea seeds and place them in the 'filter' tubes (Fig. 16.1). Insert the tubes into 200 ml of 95% ethanol contained in a 250 ml Erlenmeyer flask for 10 s, remove, shake off the excess ethanol and place in a second flask containing 200 ml of 1% Tween 20 solution. Shake gently for 1 min and then examine the individual seeds, noting and discarding those with translucent patches beneath the testa caused by the penetration of water through damaged areas. Seeds with an intact testa will remain unchanged and can then be surface sterilised by transferring the 'filter' tube with contents to a third flask containing 200 ml of 10% w/v calcium hypochlorite for 20 min. During the sterilisation procedure transfer the seeds in the flask to the sterile transfer room. After the sterilisation period is complete take the 'filter' tube with the seeds through three changes of sterile distilled water (200 ml in each 250 ml Erlenmeyer flask) to remove all traces of hypochlorite. Pour 20 ml of distilled water into each of eight 90 mm Petri dishes and place two pea seeds in each dish ca. 50 mm apart. Repeat until all the seeds have been processed and incubate in the dark for 48 h at 25 °C. When the primary seedling root has reached a length of 20–25 mm excise (using aseptic precautions) the terminal 10 mm and transfer two of these tips to a 90 mm Petri dish containing 20 ml of culture medium M_{ROOT}. Repeat until all of the seedlings have been used. Transfer the dishes carefully to a darkened incubator at 25 °C. Ensure that the inocula are only removed from sterile seedlings and that the liquid medium is not displaced from the Petri dishes.

Scheduling

Event	Timing
Sterilisation of pea seeds and incubation under aseptic conditions	Day 0
Removal of primary seedling root and initiation of culture	Day 2
Daily measurement of root length	Day 3–21
First appearance of lateral roots	c. Day 5–6

Fig. 16.2 Roots growing in Petri dish

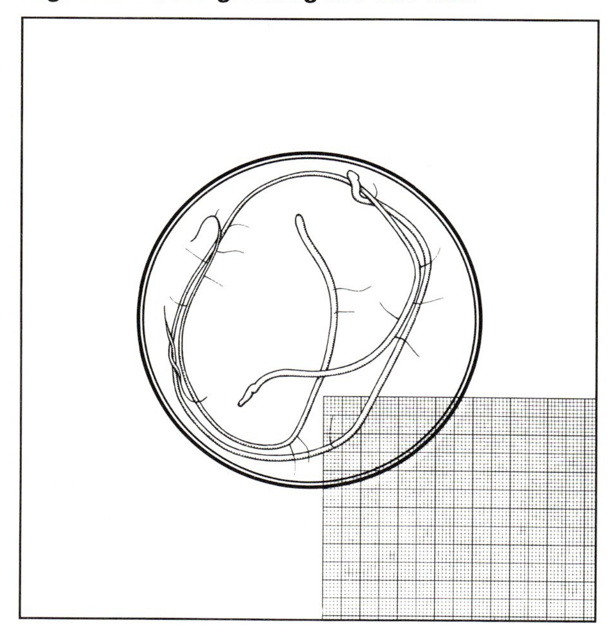

Fig. 16.3 Root growth curve

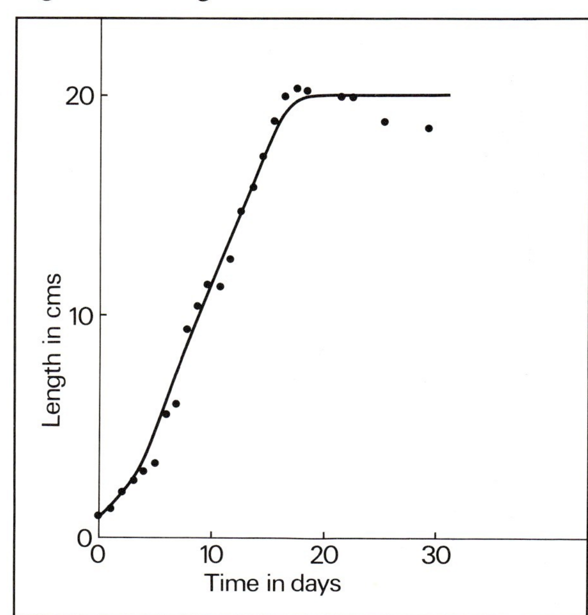

Recording of Results

Copy down the details of the experiment into a notebook recording the starting date of the experiment, the duration, number of cultures and the treatment employed. At intervals of 24 h measure the lengths of the roots by placing the Petri dishes on a sheet of mm graph paper and estimate the length against the graduations of the paper (Fig. 16.2). Plot your results graphically and analyse the growth curves obtained (Fig. 16.3).

Questions and Comments

Would the use of different lengths of the primary inoculum affect the subsequent rate of growth of the cultures?

The rate of growth is not maintained after a series of subcultures unless the medium is supplemented with thiamin and nicotinic acid. Please explain.

References

Brown, R., Wightman, F.: The influence of mature tissue on division in the meristem of the root. J. Exp. Bot. *3,* 253–263 (1952)

Kotte, W.: Wurzelmeristem in Gewebekultur. Ber. Dtsch. Bot. Ges. *40,* 269–272 (1922)

Robbins, W. J.: Cultivation of excised root tips and stem tips under sterile conditions. Bot. Gaz. *73,* 376–390 (1922)

Street, H. E., Butcher, D. N.: Excised root culture. Bot. Rev. *30,* 513–586 (1964)

White, P. R.: The Cultivation of Plant and Animal Cells, 2nd Ed. New York: Ronald Press 1963

Experiment 17 Isolation and Culture of the Shoot Apex of Carnation

Experiment 17 Isolation and Culture of the Shoot Apex of Carnation

Apices from the shoots of plants may be removed and cultured successfully under completely sterile conditions. This technique has been used increasingly over the past fifteen years to obtain virus-free plants from clones of vegetatively propagated species infected by viruses, and to build up stocks of chosen ornamental hybrid varieties or species which are slow or difficult to rear from seed. The techniques used today have been exploited by many for the propagation of dahlia, potato, carnation, narcissus, strawberry, orchids and many other plants. In this experiment carnation has been chosen as a convenient plant for the study of tip culture because it grows throughout the year, produces axilliary buds freely and, as it has smooth leaves arranged in opposite pairs, dissection of the apices is not difficult.

Materials and Equipment

Sterile Items

(For sterilisation procedure and composition of medium see Appendix)

1. 6 culture tubes (150×25 mm) containing 10 ml of M_{CAR} agar medium with the auxin NAA, sloped at an angle of 25°
2. 6 culture tubes (150×25 mm) containing 10 ml of M_{CAR} agar medium, without NAA, sloped at an angle of 25°
3. 6 Petri dishes 90 mm in diameter, glass or plastic
4. 1 pair Watchmaker's forceps
5. 1 eye scalpel
6. 12 sheets of aluminium foil, (100×100 mm)

Non-Sterile Items

7. 6 dry side shoots of carnation *(Dianthus caryophyllus)* (35–50 mm long)
8. Cotton wool
9. 1 waterproof marking pen
10. 1 low-power dissecting binocular microscope
11. 1 rack, metal or plastic, to hold 12 culture tubes (150×25 mm)
12. 1 roll of parafilm
13. 6 small plastic pots (c. 50 mm in diameter, with a drainage hole) filled with sterile potting compost, covered with a thin layer of peat and sand

Fig. 17.1 Items for the sterile transfer room

14 6 plastic pots (70 mm in diameter, with a drainage hole) filled with sterile potting compost and covered with a thin layer of peat and sand
15 6 glass beakers (100 ml)
16 1 Bunsen or ethanol burner
17 1 Erlenmeyer flask (150 ml) containing 100 ml 95% ethanol

Items for the sterile transfer room are laid out as shown in Fig. 17.1.

Experimental Procedures

Surface sterilisation of healthy, dry, carnation shoots with hypochlorite is unnecessary and appears to increase the risk of contamination. Place each shoot in a Petri dish and transfer to the bench of the sterile room. Remove the outer leaves from each shoot with a pair of jeweller's forceps. This lessens the possibility of cutting into the softer underlying tissues. After the removal of all outer leaves the apex is exposed (Fig. 17.2). Cut off the apex with the eye scalpel and transfer only those less than 1 mm in length to the surface of the agar medium with NAA. Flame the mouth of the culture tube before and after transfer of the excised tip. Close the mouth of the tube with two layers of aluminium foil and seal with a strip of parafilm to minimize evaporation. Place the tubes in the rack and incubate approximately 300 mm below 40 watt 'Natural' fluorescent tubes (continuous light) at 18 °C. As soon as root initials are visible, transfer, regardless of tip size, to tubes of fresh medium without NAA. After the plantlets have reached a height of 20 mm and have abundant roots transfer to small pots (50 mm) containing potting compost and place in a glasshouse with a minimum night temperature of 15 °C. For the first week place the plants under an inverted glass beaker (100 ml). Subsequently remove the protective vessel and expose the plant to the normal greenhouse environment. After the plants have attained a height of 50–60 mm transfer to larger pots (ca. 70 mm) containing sterile potting compost.

Scheduling

Event	Timing
Removal of the shoot tip and initiation of culture on medium *with NAA*	Day 0
Transfer of shoot tip with root initials to medium *without NAA*	c. Day 21
Transfer of plantlets (ca. 20 mm high) with abundant roots to sterile compost. Plants protected by an inverted glass beaker.	c. Day 35
Removal of the glass beaker Transfer plants to compost in larger pots when the plants have reached a height of 50–60 mm	c. Day 42

Recording of Results

Copy down the details of the experiment into a notebook recording the starting date and duration of the experiment, and the number of cultures. Make visual observations of the apices at weekly intervals recording developmental changes.

Questions and Comments

Why are you asked to culture only the apices less than 1 mm in length?

Why is it necessary to remove the auxin (NAA) immediately the root initials appear on the surface of the explant?

Fig. 17.2 Apex of carnation

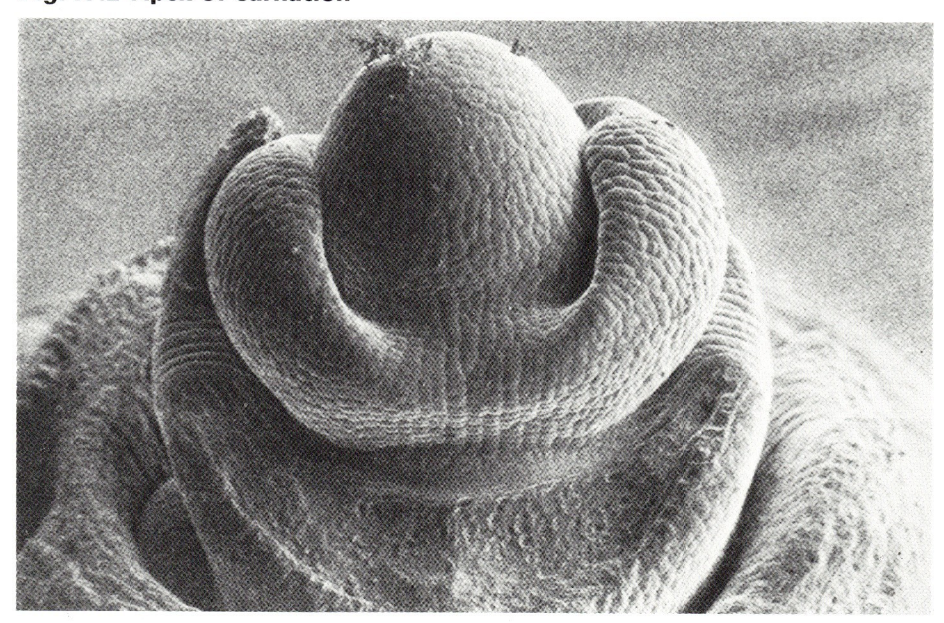

References

Baker, R., Philips, D. J.: Obtaining pathogen-free stock by shoot tip culture. Phytopathology *52,* 1242 (1962)

Boxus, Ph., Quoirin, M., Laine, J. M.: Large scale propagation of strawberry plants from tissue cultures, pp. 130–143. In: Applied and Fundamental Aspects of Plant Cell, Tissue and Organ Culture (Reinert, J., Bajaj, Y. P. S. eds.). Berlin-Heidelberg-New York: Springer 1977

Holdgate, D. P.: Propagation of ornamentals by tissue culture, pp. 18–43. In: Applied and Fundamental Aspects of Plant Cell, Tissue and Organ Culture. (Reinert, J., Bajaj, Y. P. S. eds.). Berlin-Heidelberg-New York: Springer 1977

Quak, F.: Meristem culture and virus-free plants, pp. 596–615. In: Applied and Fundamental Aspects of Plant Cell, Tissue and Organ Culture. (Reinert, J., Bajaj, Y. P. S. eds.). Berlin-Heidelberg-New York: Springer 1977

Stone. O. M.: Factors affecting the growth of carnation plants from shoot apices. Ann. Appl. Biol. *52,* 199–209 (1963)

Walkey, D. G. A.: In vitro Methods for Virus Elimination pp. 245–254. In: Frontiers of Plant Tissue Culture. (Thorpe, T. A. ed.). International Association for Plant Tissue Culture, Calgary 1978

VIII Appendix

Appendix A Selected Reading List

1 Gautheret, R.J. La Culture des Tissue Végétaux. Masson et Cie., Paris. (1959)

2 White, P.R. The Cultivation of Plant and Animal Cells. 2nd Ed. Ronald Press, New York. (1963)

3 Butenko, R.G. Plant Tissue Culture and Plant Morphogenesis. Israel Program for Scientific Translations. Jerusalem. (1968)

4 Street, H.E. Tissue Culture and Plant Science 1974. Academic Press, London, New York, San Francisco. (1974)

5 Gamborg, O.L. and Wetter, L.R. Plant Tissue Culture Methods. National Research Council of Canada. (1975)

6 Pierik, R.L.M. Plantenteelt in Kweekbuizen. Thieme & Cie, Zutphen. (1975)

7 Thomas, E. and Davey, M.R. From Single Cells to Plants. The Wykeham Science Series, Wykeham Publications (London) Ltd. (1975)

8 Butcher, D.N. and Ingram, D.S. Plant Tissue Culture. Edward Arnold. Institute of Biology, Studies in Biology. London. (1976)

9 Barz, W., Reinhard, E. and Zenk, M.H. Plant Tissue Culture and Its Bio-Technological Application. Springer-Verlag. Berlin, Heidelberg, New York. (1977)

10 Reinert, J., and Bajaj, Y.P.S. Applied and Fundamental Aspects of Plant Cell, Tissue and Organ Culture. Springer-Verlag. Berlin, Heidelberg, New York. (1977)

11 Street, H.E. Plant Tissue and Cell Culture. 2nd Ed. Botanical Monographs Volume 11. Blackwell Scientific Publications, Oxford, (1977)

12 Thorpe, T.A. Frontiers of Plant Tissue Culture 1978. International Association for Plant Tissue Culture. Calgary. (1978)

13 Sala, F., Parisi, B., Cella, R. and Ciferri, O. Plant Cell Cultures: Results and Perspectives. Elsevier/North Holland. Amsterdam, New York, Oxford. (1980)

14 Vasil, I.K. Perspectives in Plant Cell and Tissue Culture. Int. Rev. Cytology. Supps. IIA and IIB. Academic Press. New York, London (1980)

15 Staba, E.J. Plant Tissue Culture as a Source of Biochemicals. CRC Press Inc., Florida. (1980)

16 Ingram, D.S., and Helgeson, J.P. Tissue Culture Methods for Plant Pathologists. Blackwell Scientific Publications. Oxford (1980)

17 Yeoman, M.M., and Truman, D.E.S. Differentiation *In Vitro*. British Society for Cell Biology Symposium. Cambridge University Press, Cambridge. (1982)

18 Thorpe, T.A. Plant Tissue Culture. Methods and Applications in Agriculture. Academic Press Inc., New York. (1981)

Appendix B List of Commercial Suppliers

Culture Media

Becton, Dickinson (UK) Ltd. (BBL)	York House, Empire Way, GB-Wembley, Middlesex
DIFCO Laboratories (UK) Ltd.	PO Box 14B, Central Avenue, GB-East Molesey, Surrey
Flow Laboratories Limited	2nd Avenue, Industrial Estate, GB-Irvine KA12 8NB
Gibco-Biocult Limited	3 Washington Road, Sandyford Industrial Estate, GB-Paisley
Nung GmbH	Goethestrasse 5, D-6200 Wiesbaden 12
Oxoid Limited	Wade Road, Basingstoke, GB-Hants, RB24 OPW

Glassware and Plastics

Becton, Dickinson (UK) Ltd. (Falcon)	Laboratory Products Department, York House, Empire Way, GB-Wembley, Middlesex
Dynatech Laboratories Limited (Cooke)	Daux Road, GB-Billinghurst, Sussex
Flow Laboratories Limited (Linbro)	2nd Avenue, Industrial Estate, GB-Irvine KA12 8NB
Gibco-Biocult (Nunc)	3 Washington Road, Sandyford Industrial Estate, GB-Paisley
Glaswarenfabrik (Karl Mecht)	D-8741 Sondheim/Rhön
Hirschmann Laborglas	Hauptstrasse 7–15, Postfach 23, D-7101 Eberstadt
Jobling Laboratory Division (Corning)	Stone, GB-Staffs ST15 OBG
Portland Plastics Limited	Knightsbridge House, 197 Knightsbridge, GB-London, SW7
Rudolf Brand GmbH & Co.	Laborgeräte und Vakuumtechnik, Postfach 310, D-6980 Wertheim/Main 1
Sterilin Limited	12–14 Hill Rise, GB-Richmond, Surrey

Filtration

Forschungsinstitut Berghof GmbH	Abteilung Membranfilter, Postfach 1523, D-7400 Tübingen
Gelman Instruments spA	Via Lambro 23/25, 20090 Opera, I-Milan
Millipore (UK) Limited	Millipore House, Abbey Road, GB-London, NW10 7SP
Pall (Europe) Limited	Walton Road, Farlington, Portsmouth, GB-Hants, P06 1TD
Reichelt Chemietechnik GmbH & Co.	Rohrbacher Strasse 74, D-6900 Heidelberg 1
Sartorius Membranfilter GmbH	D-6450 Hanau
Schleicher & Schüll GmbH	Postfach 1, D-1335 Dassel

Incubators and Ovens

A.R. Horwell Limited (Sage Instuments)	2 Grangeway, Kilburn High Road, GB-London, NW6
Astell-Hearson Limited	172 Brownhill Road, GB-London, SE6
Baird & Tatlock (LTE)	Freshwater Road, Chadwell Heath, GB-Essex
Camlab (Glass) Limited (Cab-Line)	GB-Cambridge
Fisons Scientific Apparatus	Bishop Meadow Road, Loughborough, GB-Leicester LE11 ORE
Flow Laboratories Limited	2nd Avenue, Industrial Estate, GB-Irvine, KA12 8NB
LEEC	Private Road No. 7, GB-Colwick Estate, Notts.
Messgeräte-Werk Lauda (Incubator) Dr. R. Wobser KG	Postfach 350, D-6970 Lauda-Königshofen
Precision Scientific Co.	3737 West Cortland Street, Chicago, 111 60647, USA
Strands Scientific (National CO_2 Inc.)	Scientific House, 32 Bridge Street, Sandiacre, GB-Nottingham NG10 5BA

Laminar Flow Hoods

ASSAB Medicin Technik GmbH	Hüttenstrasse 8, D-3000 Hannover 1
John Bass Limited	Fleming Way, GB-Crawley, Sussex
Camlab	GB-Cambridge
Fell Clean Air (1971) Ltd.	10 New Road, GB-Newhaven, Sussex
Gelman Instruments spA	Via Lambro 23/25, 20090 Opera, I-Milan
Germfree Labs. Inc.	2600 SW 28th Lane, Miami, Florida, 33133, USA
Microflow Limited	The Mill, Minley Road, GB-Fleet, Hants

Cell Freezers

B.O.C. Cryo Products	B.O.C. Limited, Deer Park Road, Morden, GB-London, SW19
Comark Electronic Limited	Brookside Avenue, Rustington, GB-Sussex BN16 3LF
Cryogenic Services Limited	19 Byemoor Avenue, Great Ayton, GB-Middlesborough, Teeside
GFL Gesellschaft für Labortechnik mbH & Co.	Schulze-Delitzsch-Str. 4, D-3006 Burgwedel 1
Linde AG	Postfach 501580, D-5000 Köln 50
Minnesota Valley Eng. Inc.	New Prague, Minnesota 56701, USA
Union Carbide (UK) Limited (LINDE)	Engineering Products Division, Redworth Way, Aycliffe Industrial Estate, Nr. Darlington, GB-Co. Durham

Autoclaves

Astell-Hearson Limited	172 Brownhill Road, GB-London SE6
British Sterilizers	Roebuck, Hainault, GB-Ilford, Essex
W.C. Heraeus GmbH, Produktbereich Elektrowärme	D-6450 Hanau
Melag-Apparatebau	Geneststr. 9, D-1000 Berlin 62

Microscopes

British-American Optical Ltd. (Reichert)	266 Bath Road, GB-Slough, Bucks
Wild Heerbrugg (UK) Limited	Revenge Road, Lors Wood, GB-Chatham, Kent
E. Leitz (Instruments) Ltd.	30 Mortimer Street, London, W1N 8BB
Ernst Leitz GmbH	Postfach, D-6330 Wetzlar
The Projectina Company Limited (NIKON)	Skelmorlie, GB-Ayrshire
Carl Zeiss (Oberkochen) Ltd.	Degenhardt House, 31–36 Foley Street, GB-London, W1P 9AP
Carl Zeiss	Postfach 1369/1380, D-7082 Oberkochen

Cell Counters

Coulter Electronics	High Street South, GB-Dunstable, Beds.
Grant Instruments (Cellulose)	Barrington, GB-Cambridge

Washing Machines

John Burge (Equipment) Ltd.	35 Furze Platt Road, GB-Maidenhead, Berks.
Netzsch-Newamatic	Amselweg 2, D-8264 Waldkraiburg

Other Equipment and Supplies

A.R. Horwell Limited	2 Grangeway, Kilburn High Road, GB-London
Biotronik Wissenschaftliche Geräte GmbH	Borsigallee 22, D-6000 Frankfurt/Main
B. Braun Melsungen GmbH	Postfach 110/120, D-3508, Melsungen
Bühler Laborgeräte	Im Schelmen 11, D-7400 Tübingen-Weilheim
Büchi Laboratoriums-Technik GmbH (Laboratory Equipment)	Postfach 1268, D-7332 Eislingen/Fils
Denley Instruments Limited (Pipettes, Cans, Racks)	Bolney Cross, GB-Bolney, Sussex
C. Desaga Nachf. Erich Fecht GmbH & Co. (Chromatography)	Postfach 101969, D-6900 Heidelberg 1
Eppendorf Gerätebau Netheler & Hinz GmbH (Pipettes)	Postfach 630324, D-2000 Hamburg 63
Fisons Scientific Apparatus	Bishop Meadow Road, Loughborough, GB-Leicester LE11 ORG

Flow Laboratories Limited	2nd Avenue, Industrial Estate, GB-Irvine KA 8NB
K. & B. Grubbs Instrument GmbH & Co. KG	Froschkönigweg 15, D-4000 Düsseldorf
Heidolph-Elektro KG (Stirrer)	Starenstrasse, D-8420 Kelheim/Donau
Janke & Kunkel GmbH & Co KG (Stirrer)	IKA – Werk Staufen, Postfach 1263, D-7813 Staufen
Jencons (Scientific) Ltd. (Automatic Pipetting)	Mark Road, GB-Hemel Hempstead, Herts
Knick Elektronische Messgeräte (pH-Meter)	Beuckestr. 22, D-1000 Berlin 37
Lucknam Limited (Incubator Trays)	Labro Works, Victoria Gardens, GB-Burgess Hill, Sussex
Medical Pharmaceutical Dev. Ltd. (DECON)	Ellen Street, GB-Portslade by Sea, Sussex
Mettler-Waagen GmbH (Balances)	Postfach 110840, D-6300 Giessen
Oriel GmbH (Shakers)	Im Tiefen See 58, D-6100 Darmstadt
Wm. R. Warner (Compu-Pet)	Eastleigh, GB-Hants S05 3ZQ

Appendix C Determination of Packed Cell Volume (PCV)

The growth of a cell suspension culture can be followed by measuring a selected parameter at intervals during the growth cycle. Fresh weight, dry weight, cell number, and total protein can be used but such determinations are time consuming. A simple, rapid, and effective technique is to measure total cell mass as packed cell volume. Subsequently fresh weight and total cell number may be determined on the same sample.

Using strict aseptic precautions small volumes (up to 10 ml) of the suspension culture are removed at selected intervals and placed in a graduated, conical centrifuge tube. The tubes with contents are then centrifuged at 1,000 g for 5 min and the total volume of the cell pellet read off against the graduations on the centrifuge tube. Calculate the pellet volume as a percentage of the total volume in the tube. The pellet can then be macerated and the cell number determined. One source of error with this method is that the surface of the pellet after centrifugation may not be level and therefore an estimation may have to be made of the PCV. This technique is only accurate when applied to fine cell suspensions and when standard times and speeds of centrifugation are used.

Appendix D Determination of Cell Number in Cell Suspension Cultures

Transfer the cell pellet to a 10 ml glass or polythene container with a lid and add 2 ml of 10% HCl. Gently disturb the pellet with a glass rod to ensure adequate mixing of the cells with the acid, replace the lid and put in a deep freezer overnight (16 h). Remove the sample after the cold treatment and thaw at room temperature (c. 22 °C), then add 2 ml of 10% chromic acid (a solution of chromium trioxide in water) and leave at room temperature (c. 22 °C) for 3–5 days. Before counting, pour the mixture into a 10 ml measuring cylinder and make up to a known volume with a 1:1 mixture of 10% HCl and 10% chromic acid. Return the macerate to the tube and convert the mixture to a fine cell suspension by drawing the fluid into a Pasteur pipette or hypodermic syringe and expelling it four or five times. A final treatment using a microid shaker or a 'whirlymix' completes the maceration. Further dilution of the cell suspension may be necessary if the cell count per field exceeds 250. Load the haemocytometer in the normal way ensuring even distribution of cells on the grid. Count and average six complete grids and calculate the total number according to the following formula.

$$\frac{\text{Volume of Macerate}}{\text{Volume above Grid}} \times \frac{\text{Cell Count}}{1} = \begin{array}{l}\text{Total Cell Number}\\ \text{in Pellet.}\end{array}$$

Appendix E Determination of Cell Number in Explants

Place five artichoke explants (potato tuber or carrot root may also be macerated in this way) in a 10 ml glass or polythene container with a lid, add 2 ml of 5 % chromic acid (a solution of chromium trioxide in water) and store in a refrigerator. After at least 24 h and not more than seven days remove the sample and carefully break up the explants with a glass rod. The partially disintegrated tissue may now be reduced to single cells or small cell groups by repeatedly drawing the suspension into a 5 ml hypodermic syringe or a Pasteur pipette and expelling it rapidly. After ten cycles of treatment a drop of the suspension is placed beneath the coverslip of a haemocytometer slide. Further dilution of the sample may be necessary if the count per field exceeds 250. It is very important to ensure that the suspension is homogeneous to ensure even distribution of cells on the grid. Count and average six complete grids and calculate the total cell number per explant according to following formula:

$$\frac{\text{Volume of Macerate}}{\text{Volume above the Grid}} \times \frac{\text{Average Cell Count}}{\text{Number of Explants}}$$

$$= \text{Total Cell Number per Explant}$$

Appendix F Sterilisation

Sterilise metal instruments, glassware and aluminium foil after wrapping in aluminium foil and subjecting to 150 °C for 3 h in a hot air oven. Sterilise culture media, distilled water and other stable mixtures in glass containers closed with cotton wool plugs and capped with aluminium foil in an autoclave at a pressure of 15 PSI for 15 min. Paper towels and tissues should be wrapped in aluminium foil and autoclaved. Solutions of substances which may be decomposed by autoclaving, e.g., IAA, can be filter sterilised using a Swinney adaptor and a Millipore filter of a porosity which excludes bacteria, and added to the culture medium after autoclaving. Glassware can also be sterilised satisfactorily by autoclaving.

Appendix G Cleaning of Glassware

It is extremely important to use clean glassware for culture purposes. the following recipe ensures that the surface of culture flasks, tubes and other glassware is maintained in a smooth condition.

1 Remove any surface debris from the glass vessel by brushing with a test-tube brush in running tap water
2 Boil all glassware for 30 min in a cleaning solution containing 0.80 g of sodium metasilicate and 0.90 g of Calgon per litre
3 Rinse thoroughly in tap water
4 Soak in 0.01 N HCl for 4 h
5 Rinse thoroughly in tap water
6 Soak in distilled water for at least 3 h
7 Rinse in distilled water and dry in a hot air oven

Glassware may also be washed in diluted detergent solution (e.g., Teepol), rinsed in tap water and finally in distilled water before drying in a hot air oven.

Appendix H Culture Media

It is extremely important to work only with high quality media prepared from pure ingredients. Many individual weighings are necessary for even the simplest medium and these must be made with precision. The omission of one constituent or the addition of the incorrect amount of another can have disastrous and often expensive consequences. A simple and perhaps obvious solution to these problems is to purchase prepared media in which reliability is assured (for example, Flow Laboratories). However, ready prepared media are only available to specific formulations (e.g., Murashige and Skoog, M and S) and if a medium is required which is not commercially available then you must prepare it from basic ingredients. You will discover that for many routine plant tissue culture procedures the range of standard media available will suffice, they are much more convenient and can be cheaper!

Preparation of Media from Basic Ingredients

In order to reduce the time taken to weigh out individual ingredients each time a medium is required it is convenient to prepare concentrated stock solutions of mixtures of selected components of the media and store them frozen in the deep freezer. This is only possible of course if the mixture of components is stable and does not precipitate out under the storage conditions. Substances which are not stable in a frozen state must be freshly prepared each time the medium is required and added to the final mixture of stock solutions. An example of this procedure is given for M_{SOY} (Experiment 5):

Dissolve the macroelements in 200 ml distilled water and make up to 500 ml with distilled water (solution A). Repeat with the microelements (solution B) and mix solution A (500 ml) with solution B (500 ml). This mixture can be stored frozen in a plastic container in the deep freezer. Solutions C, (Na Fe EDTA), D (NAA) and E (Plant organics) can also be stored frozen separately. To make up the culture medium add 10 ml of the AB mixture to 50 ml of distilled water, then add 1 ml of solution C, 1 ml of solution D, 0.1 ml of solution E and 3 g of sucrose; finally make up the volume to 100 ml with distilled water and adjust the pH to 5.8 with N KOH. If the medium is to be solidified then 1 g of powdered agar must be added and the mixture autoclaved under standard conditions (see Appendix F).

Stock Solutions for M_{SOY}

Solution A; Macroelements

Constituent:	Concentration g/500 ml:
KH_2PO_4	3
KNO_3	10
NH_4NO_3	10
$Ca(NO_3)_2 \cdot 4H_2O$	5
$MgSO_4 \cdot 7H_2O$	0.72
KCl	0.65

Solution B; Microelements

Constituent:	Concentration mg/500 ml:
$MnSO_4 \cdot H_2O$	106
$ZnSO_4 \cdot 7H_2O$	38
H_3BO_3	16
$Cu(NO_3)_2 \cdot 3H_2O$	3.5
KI	7.5
$(NH_4)_6Mo_7O_{24} \cdot 4H_2O$	1

Solution C; Chelated Iron

Constituent:	Concentration mg/100 ml:
NaFe EDTA	132

Solution D; Auxin

Constituent:	Concentration mg/100 ml:
NAA	20

Solution E; Plant Organics

Constituent:	Concentration mg/10 ml:
Myo-Inositol	1,000
Nicotinic Acid	5
Pyridoxin-HCl	1
Thiamin-HCl	1

Table A.1 (overleaf) Composition of Culture Media

The composition of the various culture media described in the text are presented in the accompanying Table. The media have been named so that they may more easily be identified with the individual experiments. Several of the media have the same composition and closely approximate to commercially available products, e.g., M_{ANTH}, M_{DAUC}, M_{CYM}, M_{DIO} and M_{TOB} are M and S, and differ only slightly from commercially available (e.g., Flow) MS which can be substituted with no significant effect on performance. M_{HAP} closely approximates to White's Medium (Flow). All figures are *either* in mg/litre or percentages of w/v. Concentrations of growth substances are given in the text where more than one treatment is employed.
M_{PROT} is not stable to autoclaving and must be filter sterilised.

	M_{EMB}	M_{ROOT}	M_{ART}	M_{ANTH}	M_{DAUC}	M_{HAP}
Macroelements						
$Ca(H_2PO_4)_2 \cdot H_2O$	—	—	—	—	—	—
$CaCl_2 \cdot 2H_2O$	—	—	—	440.0	440.0	—
$CaCl_2 \cdot 6H_2O$	88.8	—	—	—	—	—
$Ca(NO_3)_2 \cdot 4H_2O$	—	236.0	236.0	—	—	288.0
$CaSO_4 \cdot 2H_2O$	27.2	—	—	—	—	—
FeNa EDTA	—	—	—	—	—	—
$FeSO_4 \cdot 7H_2O$	2.0	2.0	2.0	27.8	27.8	—
$Fe_2(SO_4)_3$	—	—	—	—	—	2.6
KCl	29.8	65.0	65.0	—	—	65.0
KH_2PO_4	81.6	12.0	12.0	170.0	170.0	—
KNO_3	—	81.0	81.0	1,900.0	1,900.0	80.0
$MgSO_4 \cdot 7H_2O$	48.0	36.0	36.0	370.0	370.0	737.5
Na_2 EDTA	—	—	—	37.3	37.3	—
Na_2SO_4	—	—	—	—	—	200.0
$NaH_2PO_4 \cdot 2H_2O$	—	—	—	—	—	19.0
NH_4NO_3	—	—	—	1,650.0	1,650.0	—
$(NH_4)_2SO_4$	—	—	—	—	—	—
Microelements						
$AlCl_3$	—	—	—	—	—	—
$CoCl_2 \cdot 6H_2O$	—	—	—	0.025	0.025	—
$CuSO_4 \cdot 5H_2O$	0.04	—	—	0.025	0.025	—
$Cu(NO_3)_2 \cdot 3H_2O$	—	—	—	—	—	—
$FeCl_2 \cdot 6H_2O$	—	—	—	—	—	—
H_3BO_3	0.04	—	—	6.2	6.2	1.5
KI	0.04	—	—	0.83	0.83	0.75
$MnSO_4 \cdot 4H_2O$	2.0	—	—	22.3	22.3	6.6
$Na_2MoO_4 \cdot 2H_2O$	—	—	—	0.25	0.25	—
$NH_4MoO_4 \cdot 2H_2O$	0.04	—	—	—	—	—
$(NH_4)_6Mo_7O_{24} \cdot 4H_2O$	—	—	—	—	—	—
$NiCl_2 \cdot 6H_2O$	—	—	—	—	—	—
$ZnSO_4 \cdot 7H_2O$	0.04	—	—	8.6	8.6	2.7
Sequestren 330	—	—	—	—	—	—
Sucrose	4%	4%	4%	4%	2%	2%
Glucose	—	—	—	—	—	—
Mannitol	—	—	—	—	—	—
Biotin	—	—	—	—	—	—
α **Pantothenate**	—	—	—	—	—	—
Inositol	—	—	—	—	—	—
Nicotinic Acid	1.0	—	—	0.5	0.50	0.50
Pyridoxine	0.2	—	—	0.1	0.1	0.1
Thiamine HCl	0.2	—	—	0.1	0.1	0.1
Adenine	—	—	—	—	—	—
Cysteine	—	—	—	—	—	—
Glycine	—	—	—	3.0	3.0	3.0
2,4-D	—	—	0.22	—	See Text	See Text
I.A.A.	—	—	—	—	—	—
N.A.A.	—	—	—	—	—	—
6.BAP	—	—	—	—	—	—
Kinetin	—	—	—	—	See Text	—
Casein Hydrolysate	500	—	—	—	—	—
Agar	1%	—	—	0.8%	0.8%	0.8%
pH	5.8	5.8	5.8	5.8	5.8	5.8

M_{CAR}	M_{CYM}	M_{PROT}	CPW	M_{D10}	M_{SOY}	M_{TOB}
—	—	100.0	—	—	—	—
—	440.0	450.0	1,480.0	440.0	—	440.0
—	—	—	—	—	—	—
500.0	—	—	—	—	500	—
—	—	—	—	—	—	—
—	—	—	—	36.7	13.2	—
—	27.8	—	—	—	—	27.8
—	—	—	—	—	—	—
—	—	—	—	—	65.0	—
125.0	170.0	—	27.2	170.0	300.0	170.0
125.0	1,900.0	2,500.0	101.0	1,900.0	1,000.0	1,900.0
125,0	370.0	250.0	246	370.0	71.5	370.0
—	37.3	—	—	—	—	37.3
—	—	—	—	—	—	—
—	—	170.0	—	—	—	—
—	1,650.0	—	—	1,650.0	1,000.0	1,650.0
—	—	134.0	—	—	—	—
3×10^{-5}	—	—	—	—	—	—
—	0.025	0.025	—	0.025	—	0.025
3×10^{-5}	0.025	0.025	0.025	0.025	—	0.025
—	—	—	—	—	0.35	—
10^{-3}	—	—	—	—	—	—
10^{-3}	6.2	3.0	—	6.2	1.6	6.2
10^{-5}	0.83	0.75	0.16	0.83	0.75	0.83
10^{-3}	22.3	13.2	—	22.3	14.0	22.3
—	0.25	0.25	—	0.25	—	0.25
—	—	—	—	—	—	—
—	—	—	—	—	0.1	—
3×10^{-5}	—	—	—	—	—	—
10^{-3}	8.6	2.0	—	8.6	3.8	8.6
—	—	28.0	—	—	—	—
2%	2%	1%	—	3%	3%	2%
—	—	18,000.0	—	—	—	—
—	—	100,000.0	—	—	—	—
10^{-5}	—	—	—	—	—	—
10^{-3}	—	—	—	—	—	—
10^{-1}	—	100.0	—	100.0	100.0	—
10^{-3}	0.5	1.0	—	0.50	0.50	0.50
10^{-3}	0.1	1.0	—	0.50	0.1	0.1
—	0.1	10.0	—	0.10	0.1	0.1
5.0	—	—	—	—	—	—
10.0	—	—	—	—	—	—
—	3.0	—	—	2.0	—	3.0
—	—	0.1	—	8.84	—	—
—	—	—	—	—	—	See Text
1.0	—	1.0	—	—	See Text	—
—	—	1.0	—	—	—	—
—	—	—	—	—	See Text	See Text
0.5	—	—	—	—	—	—
0.8%	0.8%	—	—	1%	1%	0.8%
5.8	5.8	5.8	5.8	5.8	5.8	5.8

Index

T.C. Moore

Biochemistry and Physiology of Plant Hormones

1979. 164 figures, 13 tables. XII, 274 pages
ISBN 3-540-90401-8

Biochemistry and Physiology of Plant Hormones is a comprehensive account of hormonal regulation of growth and seed plant development. The author summarizes current fundamental knowledge regarding the major kinds of hormones and the phytochrome pigment system, reflecting the steady output of important new discoveries in the field. Chapter 1 introduces the reader to the growth and development of whole plants throughout ontogeny. This sets the stage for a consideration of hormonal regulation, specifically where it concerns auxins, gibberellins, cytokinins, abscisic acid and related compounds, ethylene, and phytochrome. Biochemical aspects of hormonal regulation are emphasized throughout the book.
Biochemistry and Physiology of Plant Hormones will be a valuable text and major reference for advanced students as well as researchers in biology, botany, and such fields of applied botany as agronomy, forestry, and horticulture.

T.C. Moore

Research Experiences in Plant Physiology

A Laboratory Manual

2nd edition 1981. 23 figures. XIV, 348 pages
ISBN 3-540-90606-1

From the reviews: "This laboratory manual is one of the very best I have seen. Although it is specifically for use in plant physiology, it has the qualities of organization, instructions, references, and content which instructors and students look for in a laboratory manual for any subject area. Each experiment has an introduction of the topic area in general and leads the student into basic problems and exercises. In almost all cases the open-endedness of the topic is revealed, thus allowing extensive experimentation by advanced students. ... This feature is one of several which make the manual useful for most levels of plant physiology from elementary to advanced. ... Each exercise has complete methods and materials sections. Concise and precise explanations of procedures are included along with the details of how to use specific instrumentation... an excellent choice for teaching both undergraduate and graduate laboratories in plant physiology."
The American Biology Teacher

Plant Growth Substances 1979

Proceedings of the 10th International Conference on Plant Growth Substances, Madison, Wisconsin, July 22–26, 1979
Editor: F. Skoog
1980. 209 figures, 62 tables. XVI, 527 pages
(Proceedings in Life Sciences)
ISBN 3-540-10182-9

This volume in the Proceedings in Life Sciences Series adheres to the high standards of scientific excellence, covering the full spectrum of research in this rapidly growing field. The proceedings of two previous conferences on plant growth substances - also published by Springer-Verlag - met with great critical acclaim. Significant work by leading international authorities has been included. The 50 papers present vital information on plant growth substances and hormonal regulation. This volume, with its wealth of data, will prove useful not only to all plant physiologists, but to botanists in general.

Applied and Fundamental Aspects of Plant Cell, Tissue and Organ Culture

Editors: J. Reinert, Y.P.S. Bajaj
1977. 181 figures. XVI, 803 pages
ISBN 3-540-07677-8

Springer-Verlag
Berlin Heidelberg New York